Energy and Climate Change

Energy and Climate Change

Europe at the Cross Roads

DAVID BUCHAN

Published by the Oxford University Press
for the Oxford Institute for Energy Studies
2009

OXFORD

UNIVERSITY PRESS

Great Clarendon Street, Oxford OX2 6DP

Oxford University Press is a department of the University of Oxford.
It furthers the University's objective of excellence in research, scholarship
and education by publishing worldwide in

Oxford New York

Auckland Cape Town Dar es Salaam Hong Kong Karachi
Kuala Lumpur Madrid Melbourne Mexico City Nairobi
New Delhi Shanghai Taipei Toronto

with offices in

Argentina Austria Brazil Chile Czech Republic France Greece
Guatemala Hungary Italy Japan Poland Portugal Singapore
South Korea Switzerland Thailand Turkey Ukraine Vietnam

Oxford is a registered trade mark of Oxford University Press
in the UK and in certain other countries

Published in the United States
by Oxford University Press Inc., New York

© Oxford Institute for Energy Studies 2009

The moral rights of the author have been asserted
Database right Oxford Institute for Energy Studies (maker)

First published 2009

British Library Cataloguing in Publication Data

Data available

Library of Congress Cataloguing in Publication Data

Data available

Cover designed by Ingram Pinn
Typeset by Philip Armstrong, Sheffield
Printed by Hobbs the Printers Ltd, Totton, Hampshire

ISBN 978-0-19-956990-8

1 3 5 7 9 10 8 6 4 2

To Lisa who was writing about energy long before I started

CONTENTS

FIGURES

MAPS

TABLES

KEY TO COUNTRY NAMES, GLOSSARY OF ACRONYMS AND OF TECHNICAL TERMS

Key to tables

A number of tables in this book come from European Commission publications. Many of these tables use the official EU two-letter abbreviations for the 27 member states, which are listed below.

Belgium	BE	Luxembourg	LU
Bulgaria	BG	Hungary	HU
Czech Republic	CZ	Malta	MT
Denmark	DK	Netherlands	NL
Germany	DE	Austria	AT
Estonia	EE	Poland	PL
Ireland	IE	Portugal	PT
Greece	EL	Romania	RO
Spain	ES	Slovenia	SI
France	FR	Slovakia	SK
Italy	IT	Finland	FI
Cyprus	CY	Sweden	SE
Latvia	LV	United Kingdom	UK
Lithuania	LT		

Acronyms

DG Comp	Directorate General for Competition.
DG Tren	Directorate General for Transport and Energy.
ECT	Energy Charter Treaty
ERGEG	European Regulators' Group for Electricity and Gas
IEA	International Energy Agency
ISO	Independent System Operator

OECD	Organisation for Economic Cooperation and Development
OU	ownership unbundling.
LNG	Liquefied Natural Gas.
TSO	Transmission System Operator

Technical terms

Third party access. This effectively means open access for all. The term third (or fourth, fifth etc) party is used to indicate additional individuals or companies beyond the original partners or participants in a project, network or transaction (who would be the first and second parties)

Balancing mechanisms. Supply and demand need to be balanced in order to prevent electricity grids from crashing and to maintain steady pressure in gas pipelines. Storage helps do this in gas, but is impossible in electricity. Pricing mechanisms penalise those who cause an imbalance and reward those who can rectify it. Balancing requirements and mechanisms can be abused, sometimes by dominant suppliers and often to the disadvantage of new entrants such as wind generators injecting intermittent power into the grid.

Bcm – billion cubic metres, common measure of gas, usually the flow per year.

Gigatonnes (as in tonnes of carbon) – 10 to the power of 9, i.e. 1 billion tonnes.

Megawatts (measurement of electricity) – 1,000 kilowatts or 1m watts.

ACKNOWLEDGEMENTS

I am very grateful to the director and fellow members of the Oxford Institute for Energy Studies for giving me the opportunity to write this book, and making it all the better for their comments. In this regard, I thank in particular Christopher Allsopp, the institute's director, Malcolm Keay, Charles Henderson and Robert Ritz. For doing such a nice job of turning text into tome, my gratitude goes to Frances Kennett, Phil Armstrong, Kate Teasdale, and to Ingram Pinn for his remarkable penmanship on the cover. I have also greatly benefited from the willingness of colleagues at Argus Media to share some of their unrivalled knowledge of energy markets.

Needless to say, this book could not have been written without the help of the many EU and national government officials, industry executives and independent experts who were willing to share with me the information and insights necessary to build up a complex picture. Since they usually spoke on condition of anonymity, I must thank them collectively.

David Buchan
January 2009

CHAPTER 1

TAKE-OFF

Surely, this is just the moment in history for which the European Union was created.

Prince Charles talking on climate change to the European Parliament, February 2008.

Climate change is transforming energy policy in the European Union. The scale of the problem – the risk of irreversible warming from the world overdosing on fossil fuels – dwarfs Europe's more traditional preoccupations with energy market structures and stable supply. The international nature of the required response to global warming has put the EU institutions centre-stage, too. EU member states automatically look to their Union for solutions to the ultimate cross-border problem, in a way that they have never done with other aspects of energy policy.

From the dawning of popular awareness about global warming, EU member states have sought a collective solution. Two years before the United Nations earth summit in Rio de Janeiro in 1992, EU ministers were talking of stabilizing emissions 'in the Community as a whole'. The nature of global pollution makes solutions even on the scale of the EU insufficient. But it is politically significant that on no other issue in the EU's 50-year history – not currency, not defence, not foreign policy, not agriculture – have member states become so unanimous on the need for EU-level action.

This consensus may, of course, still depend on what the action is. In 1991 the European Commission proposed that the action should take the form of an EU energy-cum-carbon tax (to curb carbon emissions). This tax was strongly opposed by some member states, and might be again if the idea were re-proposed. The EU has since settled on a more disguised – therefore more palatable – form of carbon taxation, dressed up as the cost of carbon allowances traded on Europe's Emission Trading Scheme (ETS). Still, this system imposes costs, which seem very unwelcome in the light of the banking crash of autumn 2008 and the consequent economic slowdown. Not surprisingly, the only way the EU could agree, in December 2008, on its new climate change programme for 2013–20 was by continuing to allow companies in various countries and in various sectors to get a sizeable share of allowances for free.

Yet the political mood about climate change in 2008 is very different from 1991. The rate of increase in greenhouse gases – two-thirds of which come from energy use – has visibly and worryingly accelerated. Cost/benefit calculations of environmental action have altered, not least because of the 2006 report on the economics of climate change by Professor Nicolas Stern. His description of climate change as the biggest 'market failure' ever is no way invalidated by the autumn 2008 bank crashes.

In the EU, climate change has developed an integrationist dynamic that has only been paralleled by the '1992 single market' programme. This dynamic has shown itself in the way EU climate change policy touches on the core of nations' sovereignty by dictating the renewable share of their energy mix; in the way that parts of national policy-making are moving over to the EU and its ETS; and in the way that one piece of legislation always seems to require another. And almost no member state complains about the principle of this, though many complain about the practice. Governments may object to the precise measure of their renewable target, but not to the setting of these targets at EU level or by EU agreement. A government like the UK's is very concerned about the stability of the ETS because its environmental policy now relies so heavily on this European instrument. The EU sets overall emission reduction targets that then require further legislation to cut fuel use in cars or increase insulation in buildings. But there are few objections to this domino effect of one piece of legislation leading to another.

By contrast, national resistance to EU prescriptions is still strong in the two other main areas of EU energy concern: the drive to liberalize and integrate Europe's energy markets and the effort to improve the security of energy supplies from countries such as Russia.

It is not surprising that some member states, with a big industrial base at stake such as Germany, or with chronic competitiveness worries such as Italy, or poorer ones as in Central and Eastern Europe, are nervous at taking climate change measures ahead of much of the rest of the world. But EU members ought to feel they can be bolder as a bloc than as individual states. For countries worried about saddling their industry with environmental costs, there is economic safety in numbers. Being part of a 27-country regional trading bloc means that many EU states do most of their commerce with each other. (Contrast this with Canada's predicament – its partial backtracking on the Kyoto Protocol has a lot to do with its wide exposure to trade with the US, which opted out of Kyoto).

If the US had stayed in Kyoto, the EU might not have occupied the driving seat in global climate change negotiations. But its leadership became inevitable once the Bush administration pulled the US out in 2001. The EU was very instrumental in getting the Kyoto Protocol signed. It was central to cajoling Russia into the key ratification that brought the Protocol into effect. It set up the ETS, Kyoto's main implementation mechanism. And it is leading the way for the next phase with a programme to extend emission reductions with a 20 percent cut by 2020 (from 1990 levels) and a promise of a 30 percent cut, if others were to match it.

Less obviously, the EU is also offering a possible model to the rest of the world about how richer and poorer countries can combine to carry out differentiated emission reductions. The EU's particular clout in international negotiations stems from its states' willingness to have their greenhouse gas emissions treated as a single 'bubble' and to take on a collective commitment to reduce it. Delivering on this commitment requires hard internal bargaining about which countries do what to reduce the overall bubble.

But the EU did it with Kyoto. It came to a burden sharing agreement that at one extreme required rich Luxembourg to cut its emissions by 28 percent and at the other allowed poorer Portugal to continue its catch-up growth with a 27 percent increase in emissions. For emissions reductions after Kyoto, the burden-sharing negotiations become more complex because the EU has nearly twice the membership it had at the time of Kyoto's signing. But for emissions outside the ETS (60 percent of the total), the EU has agreed differentiation ranging from a 20 percent increase for Bulgaria to a 20 percent decrease for Denmark and Ireland.

Naturally, in a permanent Union of neighbouring countries that aspire to common values and have common policies to promote economic cohesion, richer states would be expected to take on more of the burden than poorer members. Even so, it was clear in autumn 2008 that getting agreement on a system of differentiated responsibilities and actions between richer west European and poorer east European states would not be easy. But now that the EU has got such an agreement among its members – in a Union where the spread in wealth per capita between richest and poorest (Luxembourg and Bulgaria) is greater than the income gap between the US and China – it should prove an example to the rest of the world.

But getting agreement is only part of the battle, ensuring it works is the other. The EU will only be a convincing example to the rest of the world if its climate change measures convince. It is unfortunate that a general economic slowdown should loom just as the EU was working up a comprehensive climate programme for 2013–20. It would be doubly unfortunate if a temporary slowdown, which will surely be over by 2012, became an excuse to weaken a much longer programme, and there are some fears that this is what happened in the December 2008 agreement. Europe cannot afford to waste the next decade. This is not an experiment that, if it fails, can be re-run without the underlying global warming problem being, by then, worse.

But concern about climate change – the focus of Chapters 10–12 in this book – does not sweep aside the two more traditional strands to EU energy policy that are dealt with in earlier chapters – market liberalization and security of supply. Indeed the EU has never pursued one strand of energy policy to the exclusion of the other two. The three strands or threads of energy policy have always been interwoven. For energy policy actually combines economic policy, security policy and environmental policy. The emphasis in policy changes according to events – rather like a skier, shifting his weight from ski to ski to turn, although if climate change is the avalanche behind him straight downhill might be wiser.

These shifts of emphasis create a sort of timeline, which the book follows. So the first set of issues to be discussed focus on market liberalization (Chapters 3–7), because it has been the most active area of EU energy policy-making in recent years. It started as a postscript to the 1992 single market programme, then gathered momentum as the EU developed the argument that efficient energy needed to be a building block of its so-called 'Lisbon agenda' ambition to make Europe the world's most competitive economy. But the European Commission has been trying to give liberalization an ultimate push – further than anywhere else in the world – with proposed legislation to take networks out of vertically integrated energy companies. This has led to a heated debate over the economic pros and cons of unbundling in terms of investment, price, market structure and transparency. The confrontation over ownership unbundling analysed in Chapter 6 and charted in Chapter 7 also came just when energy insecurity had made some governments (see Chapter 5) very loath to allow Brussels touch their national energy champions. Climate change has not eclipsed liberalization policies, but has significantly altered them with the introduction of non-market mechanisms like targets and subsidies for the generation of more renewable energy.

The other traditional preoccupation of EU policy is security of supply. This is the oldest common energy concern of EU states, going back to the 1973 Arab oil boycott, and arguably to the EU's very beginnings. But the concern returned with a jolt in January 2006 when Gazprom sharply reduced gas flowing through Ukraine, conduit for 85% of Russian gas reaching the EU, as part of what has become a chronic payment dispute with Kiev that has serious knock-on effects on some EU members. The jolt became a shock in January 2009 when Gazprom stopped stopped for two weeks all gas supply to and through Ukraine. Repeated scares about cut-offs of Russian gas have fuelled intense debate in the EU about how to respond to Russia's erratic behaviour towards transit countries (which never arose when these countries were all part of the Soviet Union)

and to Russia's reassertion of state control over its own hydro-
carbon resources. These issues are addressed in Chapter 9.

Any EU response, however, is complicated by the tension
between, on the one hand, smaller east European EU states,
which clamour for the EU to speak with a common voice on
their collective behalf to outside suppliers like Moscow, and,
on the other hand, bigger states which are still content to settle
energy ties with Russia bilaterally. In the past, EU authorities
have had little legal right to involve themselves in securing energy supply. Only in recent years have governments been ready
to confer on their Union formal competence to 'ensure security
of energy supply'. And even this must wait upon decisions about
implementing the Treaty of Lisbon, after the Irish voted it down
in their June 2008 referendum.

This book, while following the time line of the shifts in EU
energy policy, does not attempt to be a history of it, certainly
not before 1990. But three points need to be made about the
evolution and distinctiveness of EU energy policy.

A misleading debut

First, in order to appreciate the distance travelled in policy
terms, it is important to realize there never was, at the outset
of the EU, a golden age for energy policy-making. The early
institutions appeared to give energy great prominence, yet had
less to them than met the eye. The 1951 European Coal and
Steel Community treaty was the essential building block for
the creation of the European Economic Community (EEC).
It did not, however, lead to a common energy policy covering
energy sources other than coal. It became essentially a social
instrument to assist, with money, the run-down of west European
coal mining. That job largely done, the treaty was allowed to
expire in 2002. The 1957 Euratom treaty still exists, but it had
to wait until 2002 to get a European Court of Justice ruling
(against member states' challenge) that the Union, as Euratom,
had legal competence in the vital area of reactor safety. The EU
contribution to developing nuclear power in Europe has been
less than stellar (see Chapter 13).

Moreover, neither nuclear power nor coal proved to be the expanding energy source that oil emerged to be. And when the supply crisis in oil first came, in 1973–4, it was dealt with at a wider level by the founding, on American initiative, of the International Energy Agency to organize emergency oil stocks among its members that include some but not all EU countries.

Organic growth

Second, over subsequent years the EU dimension of energy policy grew organically. Where necessary, policy-makers borrowed legal competence from the economic and environmental parts of the EU treaties in order to justify proposing and passing energy measures. Energy's economic importance gained recognition in the 1986 Single European Act and in the subsequent single market programme. The 1992 Maastricht treaty increased the EU's ability to act on the environment as well as giving it competence to improve cross-border energy infrastructure in a programme known as Trans–European Networks.

Undismayed by its failure to exploit the general integrationist surge around the time of Maastricht by inserting an energy chapter into the treaty, the Commission kept returning to the subject of energy policy. In 1995, 2000 and 2006 it produced 'green papers' that emphasized the need for a joined up energy policy, linking market liberalization, security of supply and increasingly climate change.

One reason, among many, why this steady thud of green papers began to have an impact was a switch of tack by the UK. As a country whose entry in 1973 into the EU coincided with major oil discoveries and development in its zone of the North Sea, the UK had traditionally been neurotic about Brussels somehow making it share oil reserves, as it had had to do with fish stocks. As long as it remained an exporter of, or even just self-sufficient in, oil and gas, the UK remained neurotic on this score. So when, in the wake of the treaties of Amsterdam and Nice that followed that of Maastricht, discussions on a further treaty (this time dubbed a constitution) started in 2002, the UK fretted for its sovereignty over North Sea reserves. The issue arose because for the first time a specific article about 'community competence' in energy was being written into the treaty. Eventually, as a constitutional convention turned into the formal intergovernmental conference that produced the 2004 draft constitution, the UK was pacified with a clause (in the energy treaty article) protecting 'a member state's right to determine conditions for exploiting its energy resources'.

But French and Dutch voters killed the constitution in their referendums of 2005, and by the time a slimmed down version of the constitution was revived in 2007 in the form of the Lisbon treaty, the UK attitude had changed considerably. Not only had the UK less to protect, as its reserves, especially in gas, began to fall steeply into deficit, but it had also become keen on having an EU policy influencing the terms under which Britain imports gas from the continent. This change – plus a growing interest in climate change – had already persuaded UK prime minister Tony Blair to make energy policy a feature of his presidency of the EU and of the Hampton Court summit in autumn 2005, which in turn helped pave the way for the Commission green paper in January 2006.

The Treaty of Lisbon, whose future Irish voters have made uncertain, took up in almost identical language the dead constitution's wording on energy. Its Article 194 stated that:

> In the context of the establishment and functioning of the internal market and with regard for the need to preserve and improve the environment, Union policy on energy shall aim, in a spirit of solidarity between member states, to:
> – ensure the functioning of the energy market;
> – ensure security of energy supply in the Union;
> – promote energy efficiency and energy saving and the development of new and renewable forms of energy;
> – promote the interconnection of energy networks.

However, there was the crucial caveat:

> Such measures shall not affect a member state's right to determine the conditions for exploiting its energy resources, its choice between different energy sources and the general structure of its energy supply.

The inclusion of security of supply as an EU goal suggested

action in an area that had always been EU energy policy's weakest link, and the solidarity reference aimed to provide reassurance to new east European member states. Basically, however, the treaty would only codify what had been happening in real life within the EU for some time. That was striking in itself. The fact that EU energy policy had developed organically, running ahead of treaty clauses, suggests that ministers and governments saw increasing value in a common policy, more perhaps than they would openly admit.

A federal energy policy unlike others

Third, the nature of this common policy is different from those in other federations. The mission statement for energy policy, cited above in the Lisbon treaty, looks broadly similar to the energy policy goals and legal competences of federal governments in such federal systems as the US and Canada. Yet it is undermined by the caveat clause, cited above, about residual national sovereignty.

So, to take internal energy, the EU dimension in energy policy is weaker than in traditional federations. In these federal systems – where fossil fuel reserves are the property of states, provinces or (in the case of the US) private landowners – the federal authorities levy royalties, impose retail taxes, and own all offshore and some onshore reserves. The EU has no power over member states' energy mix, depletion policy or taxation (at least not on upstream production). And this is what one would expect, given the generally weak centre of a very loose federation of nation states whose existence is centuries older than their 'Union'. And it leads to the obvious result that member states are totally free to pick, choose and, in the case of nuclear power, reject energy sources they dislike.

Bizarrely, however, Brussels has greater potential power to shape the energy market design of its member states than Washington has over US states or Ottawa over Canadian provinces. This stems from the market-opening and market-liberalizing provisions written into the original Treaty of Rome. It took time to apply these provisions to energy, but they now apply to energy as to every other sector of the EU economy (except, by special treaty exemption, defence).

Compare this with the situation in the US, where electricity regulation is largely left to states that are free, should they desire, to maintain monopoly utilities and to exclude out-of-state competitors. The chances, therefore, are greater of getting even a halfway standard electricity market design for the EU's 27 ancient nation states than for the 50 US states, some of which came into existence long after the American Union. (The paradox does not exist in gas. US federal regulators have powers over states in siting long distance interstate gas pipelines that the Commission only dreamt of in its Trans-European Networks programme of the early 1990s, and in its 2007 proposal for a semi-federal Agency for the Cooperation of Energy Regulators.)

Backing up the EU's market-liberalizing provisions are the Commission's powers in anti-trust enforcement and scrutiny of state aid. Exercise of these powers is subject to control by the European Court of Justice, but not by governments. This gives the Commission unusual autonomy in the anti-trust field, which as shown in Chapters 4 and 7 it has had no hesitation in using in the energy sector.

The Commission also has a unique role in controlling market-distorting state aids. The EU is alone among federations in having an executive that can stop member governments from spending their own money on their own companies in their own territory, if the result is to distort the market to the disadvantage of competitors. This power, which for obvious reasons the Commission has to exercise independently of governments, is considered an essential safeguard in a federation where state budgets vastly outweigh the central budget. For the time being, Brussels' state aid control powers are less relevant in energy, because it concedes that, in the rush to get green power projects off the ground, states can remain free to use whatever type and level of national financial subsidy they choose. However, if ever there were to be a move to a harmonized EU-wide renewable energy subsidy, Brussels' state aid powers would come into play.

Yet in external energy policy, the federal role in the EU is feeble. Only in the 2007 Lisbon treaty (if it holds) did the EU get formal competence in energy security and, by extension, external energy supply policy. There is no requirement or tradition for member states to speak with one voice when they talk about energy with outside suppliers. Likewise there is no requirement or tradition for EU governments to inform Brussels of the bilateral energy deals that they or their companies make. Few advocates of Europe 'speaking with one voice' provide a precise prescription of what they mean. Certainly the Lisbon treaty does not. Indeed the treaty could, by member states' insistence on maintaining national prerogatives in choice of energy source and supply, perpetuate the cacophony in external energy policy. The noise of 28 groups (27 plus the EU) of policy-makers all sounding off at once about energy security is confusing and invites derision. When people say dismissively: 'there's no such thing as EU energy policy', it is generally external policy they are referring to.

Such criticism now hits home because of nervousness about growing reliance on imports and an increasing belief that only a unified Europe of consumers can carry weight with foreign oil and gas producers. By 2030, Europe could be importing 84 percent of the total gas it consumes and 93 percent of its oil, according to a Commission projection.[1]

Climate change is giving a new rationale and coherence to energy policy at an EU level, because Europe is seen as the right dimension for a solution to a problem that so many want tackled. The same could go for energy security, but it hasn't so far. This could alter, depending on how the world energy scene evolves. If its main feature becomes a mad scramble for remaining fossil fuels, then the EU could be seen as a lumbering bureaucracy unable to match more agile nation states in signing up scarce supply, and as the wrong institution to be entrusted to bargain collectively on behalf of the 500 million citizens in its 27 member states. If, however, it seems possible to develop a multilateral, rules-based approach to energy security as well as to climate change, then the EU will appear to be the right institution to pursue it.

---

1   European Commission, 'An Energy Policy for Europe', COM (2007) 1, p. 3.

CHAPTER 2

TRADE-OFFS

For far too long we have been in a situation where, in a haphazard and random way, energy needs and energy priorities are simply determined in each country according to its needs, but without any sense of the collective power we could have in Europe if we were prepared to pool our energy and our resources.

Tony Blair, UK prime minister, speaking as EU Council president to the European Parliament, 2005.

If there is one characteristic that sets the enlarged EU apart from all other federations or nation states, it is diversity. And, rather aptly, this diversity can be a source of EU weakness or difficulty or strength.

Part of the energy diversity evidently lies in differing geological endowment or geographical position. Some states have a lot of coal (i.e. Germany, Poland). Some have oil and gas (UK, Netherlands, and Denmark). Some coastal states (Portugal, Spain, UK, Netherlands, Denmark, and Germany) have a lot of wind potential. Some states (in East and Central Europe) have forests of considerable biomass potential. But diversity can be man-made too. France's massive nuclear programme sets it far apart from almost all EU states in terms of reliance on atomic power for electricity generation. History also still plays a role in setting some countries apart, notably the heavy dependence of some Central European and Baltic states on Russian energy.

Such diversity can render averages meaningless for individual states. Do the Baltic countries really care that the Union is self-sufficient in electricity or only 60 percent dependent on imports for its gas, if their only outside grid connection is still with Russia, and their sole source of gas is Russian too? Here diversity is a source of weakness, at least until there are more energy interconnections among EU states.

Diversity can also be the cause of difficulty in, for instance, trying to design a common climate change policy to suit, on the one hand, France, with its 80 percent electricity dependence on low-carbon nuclear power and, on the other hand, Poland, which generates 95 percent of its power with carbon-rich coal. Provided EU states are sufficiently interlinked, however, diversity must be considered as extra security. Diversity could help ensure that if one energy source were knocked out – oil (another Arab embargo), gas (a Russian cut-off), nuclear (an accident), coal (pollution) – others could fill the gap.

Even the most diverse collection of states can gain from acting together. Collective action can produce something that is greater than the sum of the parts, as in the case of a single, trans-European energy market. It can achieve greater impact than individual countries could, as demonstrated by EU leadership in climate change negotiations. It can also be a way of limiting the costs, as well as sharing the benefits, of pursuing common energy goals, such as countries protecting their industrial competitiveness by trying to ensure that as many commercial rivals as possible shoulder the same cost of carbon. This avoids the problem of companies or countries getting a 'free rider' advantage on the backs of the collective action of others.

How and why collective EU action works better in some areas, and not at all in others, is a theme developed in this book. Below is a rough scorecard of actual EU performance, in given policy areas, set against EU potential. It does not attempt to be scientific, but rather as a spur to thinking about the EU's relevance to energy policy.

Potential is harder to judge than performance. It is more subjective, because there is little hard data. It also has to pay some regard to past history and present necessity, because these factors determine what is legally and politically possible today. If EU-level action is considered essential, there is top potential for it because sheer necessity will override politics and legalities. This is the case with climate change, where even action on an EU-wide scale is considered too geographically limited. Potential of internal market policies gets a slightly lower, but still high, rating. EU law-makers and regulators have invested a lot of time and effort in this area, and have all the necessary legislative and anti-trust tools at hand. Yet there is still some national resistance to EU energy market blueprints, and this somewhat mars the chances of success.

The potential rating for EU policies on energy security issues is lower. This is the net effect of the clear desirability for the EU to speak with one voice to its outside energy suppliers, but also its patent inability to do so because of entrenched national attitudes. There ought to be a strong EU dimension to energy research which, however, is a sector with unusually low public and private R&D investment. The EU potential in energy efficiency is somewhat reduced by the political need, on grounds of subsidiarity, to leave many decisions to national discretion and implementation. The biggest disappointment – measured as the gap between potential and performance – has been the failure of Euratom, for all its heavyweight institutional machinery, to contribute more constructively to nuclear power around the EU.

Table 1:   Buchan's Benchmark

EU policy	EU potential	EU performance
Climate change	A+	A–
Internal market	A–	B+
Security of supply	B+	D
Nuclear power	A–	D
Renewable energy	A–	C
Energy R&D	B+	C
Energy efficiency	B	C

Source: Author

However, irrespective of their rating, common policies will not suit all countries all of the time in all areas. The European Commission argues its three main policy strands – a more integrated and competitive energy market, improved energy security and climate change controls – can be woven into a seamless synthesis, and that it is perfectly possible to pursue policies to arrive at an energy supply that is competitive, secure and sustainable. A perfect equilateral triangle. And in the abstract, this may be true. In general terms, an efficient market, with cost-reflective prices that respond to carbon pricing signals from the trading system, would make maximum use of available energy resources and so keep imports and pollution as low as possible.

In practice, this is not a perfectly compatible triangle. It is possible to meet two of the three goals, perhaps any two, but policies to meet all three goals equally may be impossible, with the important exception of energy efficiency and research whose EU dimension is explored in Chapters 14 and 15. Elsewhere there will tend to be trade-offs, meaning fulfilling less of one goal to attain more of another goal.[1] Trade-offs will differ according to member state. Not all member states can achieve the three objectives equally. Nor might they want to. Left to their own devices most member states would probably come to a different balance of policy goals. Indeed it is membership of the Union that forces member states into trade-offs in energy as in other sectors.

There should be no surprise. It is part and parcel of every country's experience of the EU in which some national sovereignty is traded off for a wider benefit. So France has had to swallow far more market liberalization than it would have done, of its own volition, in order to share in a common climate change policy. Germany, left to itself as a country without an oil major of its own but with a sizeable environmental movement, would rank energy security and climate change goals above structural reform of its fiendishly complicated domestic energy market. Britain is now compromising its long-standing market liberalization by agreeing to give much more financial support to renewable energy so that it can meet its very ambitious green power target. Smaller member states, especially those in Eastern Europe with an historic mono-supply from Russia, are most concerned about energy security, but are obliged to undertake market reforms and climate change measures that interest them far less.

So member states are continuously trying to find new ways to pursue national energy preferences within the EU policy

---

1   For stimulating thinking on policy trade-offs, see Roeller, Lars-Hendrick, Juan Delgado and Hans W. Friederiszick, *Energy: Choices for Europe, 2007*, Bruegel: Brussels, 2007.

template of collective action. Another theme of this book is the Commission's constant battle to keep countries corralled within the template. And a third theme is the Commission's efforts to deal with conflicts between different EU policy goals (represented schematically in the chart below). Indeed it is ironic that Brussels should devote so much rhetoric to denying the existence of such conflicts when it spends so much time actually dealing with them.

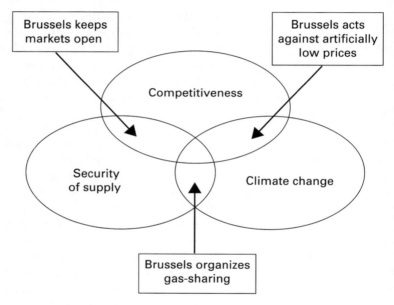

Figure 1: Reconciling Policy Conflicts

Source: Author

Let us start with the possible contradiction between competition and security of supply. As it happens, some bigger member states such as France and Germany are heavy oil and gas importers and, because of that, want to retain big national energy companies. A national champion strategy is not illogical as a deliberate attempt to create domestic market power to offset a foreign supplier's market power (such as that of Gazprom). But the creation of such national champions can have distorting consequences for the internal market (see Chapters 4 and 5). Once created, such EU national champions develop a strong affinity with non-EU national champions (such as Gazprom, see Chapter 9) in opposing structural reform inside the EU. For their part, smaller member states cannot build up national champions of any external consequence. Nonetheless, the desire to keep their incumbent energy suppliers intact leads several small states to oppose the Commission's plans for ownership unbundling (see Chapter 7).

Yet the scale of the EU helps reduce the overall loss of competition that some member states are ready to accept for the sake of security of supply. The Commission has no powers to stop mergers entirely within a single member state (see Chapter 4). But if companies get bigger or stay big, competition can still be preserved if the overall market can be enlarged. So Electricité de France may not serve French consumers very well by continuing to dominate the French market, but it renders consumers in the UK, Germany and Italy a service by invading their markets and increasing competition there. And the Commission has helped make some of these invasions possible by removing legal barriers that a few countries erected against EdF takeovers on their territory.

The interrelationship between competitiveness and the environment is somewhat different. Here, some trade-off between the two goals is probably desirable. The EU will only reach its energy efficiency and emission reduction targets if the price of fossil fuels and fossil-fuel generated electricity stays relatively high (also a goal of the emissions trading scheme). One qualification to this is the risk of 'carbon leakage' (see Chapter 10). This is the danger that, because of the cost and scarcity of EU carbon permits, EU industry will lose market share to companies outside Europe operating without any comparable carbon controls or costs, and therefore a portion of carbon 'saved' in Europe will 'reappear' elsewhere.

It is clear that attempts to keep these prices, especially of electricity, artificially low do not help the environment. One of the reasons why, for instance, Spain's emissions are way over its Kyoto target is that its politicians fix retail power prices too low (see Chapter 5). Spain is not the only country to frustrate in this way the free play of supply and demand to set prices. In theory, the EU can help here institutionally, with the Commission taking to the European Court of Justice those governments that regulate prices in defiance of EU energy directives. Naturally, such action will hardly feel like help to those governments taken to court. But it does, or should, dissuade governments from sacrificing rational energy policy to their short-term desire to please voters with energy price controls.

The Commission reacted with the same sort of advice to the 2007–08 surge in oil prices and to the clamour from fuel users for governments to make offsetting cuts in fuel taxes. The EU has no harmonized level of national excise tax on fuel, only an EU-agreed minimum floor for national fuel taxes that is well below current rates. So, in contrast to price controls that are in contravention of EU electricity and gas directives, the Commission cannot stop governments cutting their fuel taxes to the minimum to appease protestors. Nonetheless, it has highlighted the eminently sensible message to Europe's politicians that they should target financial aid directly to those hardest hit by fuel price rises, rather than cut fuel taxes in a general way that would delay the necessary shift away from fossil fuels. Obviously delivering such a message is easier for unelected Eurocrats than for politicians whose position is directly dependent on voters.

The trade-off between security of supply and the environment depends very much on a country's energy mix. At one extreme, France faces no trade-off or conflict because it largely achieves both goals through nuclear power, while at the other, Poland gains security through heavy reliance on its own coal but at the expense of pumping carbon into the atmosphere. A more collective EU energy policy could ease the security constraint felt by Poland and other central and eastern EU states that lack indigenous coal and rely solely on Russia for gas imports. This would involve Europeanizing member states' energy relations with Russia and other outside suppliers, and common arrangements for the strategic storage of gas and sharing it in emergencies (see Chapter 8). Yet while smaller member states would welcome such a common policy, the larger ones might see it as cramping their freedom to strike bilateral energy bargains with Russia and other suppliers.

All these examples are ways in which the EU tries to reconcile conflicts arising out of member states' desire to pursue national policy preferences that do not fit into the EU template. They generally involve the EU preventing member states making trade-offs. But in the climate change arena, there may be some flexible trade-offs that are *not* being made, but which perhaps *should* be. For the EU has set goals to develop more renewable energy and biofuel than is strictly necessary to meet its over-riding 20 percent greenhouse gas (GHG) reduction goal (see Chapter 10). The extra development of renewables and biofuels will add to energy security and possibly give technology a push. But they are an expensive way of avoiding carbon. So scaling down renewables and biofuels targets could help fund more ambitious cuts in GHGs, if one could be sure – and it is a very big if – that savings on renewables would be spent on clean coal or nuclear.

There are rigidities – some good, some bad – inside the proposed renewables and biofuels programmes. An example of a 'good' rigidity is the environmental criteria for biofuels (Chapter 12); without some restrictions the EU could end up using plant fuels that produce more carbon emissions than they save. An instance of a 'bad' rigidity is the likely restrictions on trade in renewables. Segmentation of the renewable market will limit the potential for scale economies, because of reliance on different national targets and subsidy schemes in individual member states, and limits on trading green power between member states (see Chapter 11).

This is a pity. For if there is one intrinsic advantage that the EU offers its members, it is that of a continental size market with economies of scale allowing producers to lower unit costs. However, many in Europe's renewable energy lobbies have a reasonable riposte to complaints about their desire to protect national subsidy schemes for green power, even at the cost of segmenting the EU market. They admit the distortions in the renewable energy market, but say policy-makers' first order of priority should be to remove the many market barriers in conventional energy that still accounts for the overwhelming share of Europe's energy consumption. And these barriers are the focus of the next few chapters.

CHAPTER 3

LIBERALIZATION: TRY, TRY AND TRY AGAIN

Energy was fundamental to the initial foundations of the EU. It has recently returned to the top of the political agenda.

Andris Piebalgs, EU energy commissioner, 2007

Given the importance energy has now assumed, it seems improbable that Europe's politicians should have left energy liberalization as a postscript to the single market blueprint launched in 1988. Yet they did so, and for a reason. Electricity and gas industries depend on networks that in turn can be considered natural monopolies. (This is in contrast to oil and coal whose physical characteristics – storability, higher energy density and flexible transport – have allowed an international market in these commodities to flourish with relatively little government intervention). Brussels left to last liberalization of these network industries – electricity and gas, but also telecommunications and parts of transport (rail) – because they all contained this element of natural monopoly on grids and pipelines and of government involvement, either in ownership or regulation.

However, in the early 1990s the Commission began to agitate against monopolies, both national and natural. National legal monopolies, generally held by state companies, had played an important part in European energy. But where they extended to legal monopoly on the import and export of energy they were clearly incompatible with a single market. And in 1991 the Commission launched a series of cases against member states that was eventually successful.[1]

At about the same time, the Commission began the search for

---

1   Peter Cameron, *Competition in Energy Markets: Law and Regulation in The European Union*, Oxford University Press, Oxford, 2007 (2nd edition), 16.51 p. 478.

a solution to natural monopolies in energy, by pushing the idea of third party access (TPA) to these grids and pipelines. TPA, which just means open access to all customers on a network, is still the spearhead principle of EU liberalization. Almost all of the other proposals emanating from Brussels are simply designed to make TPA happen. The greater powers that the Commission sought in 2007 for national regulators are largely to ensure that owners or operators of networks do not abuse their natural monopoly and give fair and equal access to other energy suppliers. The Commission's deeply controversial 2007 proposal to unbundle or separate not just the operation but also the ownership of transmission from other parts of energy businesses is just a final attempt to make TPA work, by removing once and for all the temptation for integrated energy groups to keep rivals off any networks they own.

By the mid-1990s, the Commission was making some progress, at least in the electricity sector. This eventually bore fruit in the 1996 electricity directive. This directive posed no problem to several EU states, such as the UK and Scandinavian countries that had already, through national legislation or regulation, opened up their energy markets. But it was only grudgingly accepted by France and Germany which, as we shall see, have acted in tandem to resist Brussels reshaping the structure of their energy industries.

The TPA concept was modified to allow France to keep a so-called 'single buyer' system, whereby outside suppliers could make a contract with end-user customers in France but the energy supplied had to pass through the conduit of France's 'single buyer' (at the original contract price minus network access tariffs charged by the 'single buyer'). The 'single buyer' was of course state-owned Electricité de France which was thus able to retain its lock-hold on the downstream retail market at the price of tolerating a small dose of competition from generators.

The 1996 directive was further modified to offer a choice between 'regulated' TPA, which provides access to the network on the basis of firm published tariffs approved by national regulators, and 'negotiated' TPA in which indicative prices are published, but are not binding and are not necessarily approved by regulators. Most EU member states chose the first option. But

Germany insisted on the 'negotiated TPA' option, in tune with its tradition of settling energy rules through private negotiation rather than public law. (Germany became the last EU state to set up an energy regulator, when in 2005 it established the Bundesnetzagentur which also deals with other network industries). In 1998, with the delay reflecting the somewhat greater inherent difficulties in liberalizing the gas market, a similar directive was passed 'concerning common rules for the internal market in natural gas'.

But the 1996/1998 directives did not linger long on the EU statute book. Within a few years a mood to replace them developed. One reason was the directives' unsatisfactory nature. Their timetable was ludicrously slow for market opening. For electricity, 35 percent of the market was to be open by 2003, but for gas it would be 2018 before a similar share of the market would be open to cross-border competition. Negotiated TPA proved something of a farce in Germany. Its market was theoretically 100 percent open but effectively closed to foreign suppliers who found it impossible to 'negotiate' their way in. In other states regulation of cross-border liberalization and TPA proved very weak. At the same time, buoyed up by the dotcom economic boom, EU leaders came up at their Lisbon summit in 2000 with the so-called 'Lisbon agenda' for growth and competitiveness (nothing to do with the Lisbon treaty). Accelerated energy market reform was stated to be a key component of this programme. Thus were born the two 'acceleration' directives for electricity and gas of 2003 that, in the absence of subsequent legislation, are still in force today (2008):

a) *Electricity*
The gradual exposure of customers to free cross-border choice was speeded up with full market opening for all customers set for 1 July 2007, except for a couple of smaller new member states that were given a later deadline. Only regulated TPA was allowed. Any transmission system operators (TSOs) or local distribution system operators (DSOs) that are part of larger groups must be legally unbundled, or put into separate subsidiaries. Smaller DSOs may escape this constraint, provided their operation is unbundled functionally or put under separate management. Alongside the

directive is the 2003 electricity regulation (a regulation is applied directly and in identical form in all EU states, in contrast to a directive that allows member states to translate EU legislation into their own national statutory form). Designed to foster cross-border trade, this sets out provisions on inter-TSO compensation systems, access fees and interconnectors.

b) *Gas*

This directive lays down the same market-opening timetable, by mid-2007, as the electricity directive. It also has the same un-bundling provisions – legal for all integrated TSO/DSOs and functional for smaller DSOs. But there are some modifications specific to gas. Access to storage can be restricted when TSOs need to do so in order to carry out functions related to the system or to production, but not for market purposes. TPA can be refused if it raises problems for take or pay contracts. Again the gas directive is accompanied by a 2005 regulation on rules for TPA services, and open and fair balancing (see Glossary) systems.

Still, however, there were doubts that the 2003 directives would be sufficient or sufficiently implemented. So the Commission decided to make use of a new power it had been given to launch a competition inquiry into a whole sector of the European economy without having any specific suspicions or indications of infringements. It had already done this with retail banking and telecommunications, and in June 2005 decided to do this with energy.

At the outset of this inquiry, Brussels had no 'smoking gun' in the form of proof of anti-competitive wrongdoing. But it was concerned about the 2004–5 rise in electricity and gas prices which appeared to go beyond the upward trend in other energy prices, and worried about the lack of competitive market conditions to reverse this rise. 'Cross-border flows seem insufficient to constrain price differences between most member states and integration between national markets has been slow in many regions', complained the Commission in announcing the sector inquiry. 'In addition, new market entry has been limited, and market concentration remains very high.'[2]

The 'sector inquiry', as it was known, involved Commission

2   Commission press release, IP/05/716, 13 June 2005.

officials sending out questionnaires to several thousand energy companies (which were legally obliged to respond), trawling through a massive amount of data, and carrying out some subsequent interviews. The energy expertise gained during this sector inquiry paid off in two ways. It laid the groundwork for some anti-trust cases. When the Commission opened investigations in May 2007 against Italy's Eni and Germany's RWE for allegedly shutting competitors out of their home gas markets, it claimed these investigations were not based on the sector inquiry, but were instead the result of surprise inspections carried out on company premises in May 2006. The EU competition authorities carried out surprise inspections – which the press likes to dub 'dawn raids' – on 25 companies in six states (Austria, Belgium, France, Germany, Hungary and Italy) in pursuit of allegations of foreclosure (shutting out competitors) in the gas market. In December 2006, it carried out similar inspections looking for possible abuse by companies of withholding power to manipulate prices in wholesale electricity markets.

The Commission's real reason for minimizing the sector inquiry's role in the launch of anti-trust investigations was to avoid giving the defendants an excuse to demand access to the enormous amount of general data gathered by the inquiry. Rather more truthfully, the Commission did state, at the time of the Eni and RWE inquiry announcements, that the inquiry had given it 'an in-depth understanding of the functioning, and in some cases malfunctioning, of the energy sector'. This knowledge had helped the Commission 'draw conclusions as regards where Commission investigations based on competition law could be appropriate and effective'.[3] In other words, it showed Brussels where to look, and what to look for.

More broadly, the sector inquiry provided ammunition for the Commission to make another legislative assault on the energy sector. In a 328-page report it identified 'serious shortcomings'. These are dealt with in more detail in the next chapter. The chief characteristic revealed by the inquiry was the high degree of market concentration by incumbents (generally defined as the

---

3   Commission   press   releases   Memo/07/186   percent Memo/07/187.

pre-liberalization monopolies or dominant companies) in their home states arising from two factors. One is so-called vertical foreclosure – the use by vertically integrated energy groups of their networks and long-term upstream gas contracts to shut rivals out. The other is lack of competition because of the European energy sector's continued segmentation into national markets with too few interconnecting wires and pipes to link them.

The main remedies proposed by the sector inquiry were more market transparency to help new entrants; caps on incumbent market shares and gas release programmes to increase liquidity in the market; closing the gaps between the responsibilities and competences of national regulators; and, most controversial of all, structural unbundling so as to create genuinely independent transmission system operators interested in acting as a real common carrier of energy.

This proposal of ownership unbundling was then endorsed by the full Commission, in its January 2007 package of proposed legislation, as 'the most effective means to ensure choice for energy users and to encourage investment'. The Commission said this was because 'separate network companies are not influenced by overlapping supply/generation interests as regards investment decisions. It also avoids overly detailed and complex regulation and disproportionate administrative burdens.'[4] At the same time, the Commission suggested as a second best option a 'full independent system operator where the vertically integrated company remains owner of the network assets and receives a regulated return on them, but is not responsible for their operation, maintenance or development'.

The Commission knew it was inviting trouble in proposing fresh legislation before the existing legislation was properly in place. Not only did the deadline for full market opening only come in mid-2007 according to the 2003 directives, but these current directives were not being correctly implemented. The Commission had highlighted this fact in 2006 by starting court proceedings against no fewer than 17 states for inadequate transposition of the 2003 laws. So why hurry? This was the

---

4   European Commission, *An Energy Policy for Europe*, COM (2007)1, p. 7.

Table 2:   Re-regulating Energy

	Unbundling of networks	*Access to networks*	*Market opening*	*National regulation*
First legislative package 1996–8	Separate management and accounts	Negotiated or regulated terms of access	Power: 35 percent open by 2003 Gas: 33 percent open by 2018	Mechanism for regulation
Second Legislative package 2004	Separate subsidiary	Regulated terms of access	Power and gas markets 100 percent open by July 2007	Specific regulator for energy
Third legislative package proposed 2007	Separate ownership or operator	Regulated terms of access	Already achieved (see above)	Ungraded and harmonized powers for national energy regulators

Source: Author

cry from governments like that of France and from the big incumbents. They advised Brussels to wait for all member states to implement the 2003 legislation before making any judgement. Yet, while those governments criticising Brussels for being hasty had a common sense argument – no experiment should be pronounced a failure until fully tried – most of them had undermined their case by dragging their feet on the 2003 directives.

A more important counter-argument by the Commission was that the 2003 directives were fatally flawed. As Philip Lowe, its director general for competition, told a conference in January 2007, 'the existing legislation is simply too weak to have an impact.' [5] Its unbundling provisions were too open to manipulation. Some ostensibly unbundled independent system operators

---

5   Philip Lowe and others at Claeys–Casteel conference, Brussels January 2007

(ISOs) had proved just to be shell companies with all the work contracted back by the vertically integrated parent. And in many cases, Commission officials claimed, it had proved impossible to have functional (separate management) unbundling without legal (separate subsidiary) unbundling. National regulators had also pointed out the difficulty of monitoring legal unbundling across borders when a transmission system operator in one state has links to a supplier in another country.

Furthermore, the scale of market abuse uncovered in the sector inquiry was felt to warrant new legislation across the board. Asked in January 2007 why the hurry for new legislation, one competition official who had worked on the energy sector inquiry replied incredulously: 'You mean give the companies more time after their ten years of doing nothing?'

In the end, the competition directorate's involvement was decisive to the Third package of internal market reform. To say this is not to ignore the considerable role played by Andris Piebalgs as energy commissioner 2004–9. He got off to an unpromising start to his job. He was not the first pick as commissioner by the Latvian government, whose original choice for commissioner ran into opposition in the European Parliament. Nor was he initially intended as energy commissioner by José Manuel Barroso, who switched Mr Piebalgs to this portfolio after the Hungarian candidate for this commission job performed so poorly before the European Parliament. But Mr Piebalgs took to the energy job with gusto and talent, and made a considerable success of it.

Yet, though Mr Piebalgs was to prove a doughty fighter for energy market reform, this issue was by no means the only one on his plate. So, if the Directorate-General for Competition (DG Comp) had not involved itself in energy across the board, it is likely the Third package would never have been proposed at a time, 2007, when liberalization had been overtaken as an issue by climate change and energy security. But fortuitously it did get involved. This not only stiffened the resolve in the Commission for further legislation. It also weakened governments' resistance, in the view of a senior Commission official. 'Member states realized they could either negotiate [on legislation] or face the competition directorate which they couldn't control.'

CHAPTER 4

MARKET ABUSE:
THE THINGS SOME COMPANIES DO

This requires comprehensive structural reform, [because] even the most diligent competition enforcement cannot solve all the problems in these markets.

Neelie Kroes, competition commissioner at September 2007 launch of the Commission's Third liberalization package.

You might have thought the liberalization measures mentioned in the previous chapter would have resulted in less concentration and more integration. After all, doing away with national monopolies in the import and export of energy (in the early 1990s) should have been a blow to dominant incumbents and a spur to their domestic rivals, while the phasing in of full cross-border competition (by mid-2007) should have brought foreign rivals into national markets and increased EU-wide integration. This has not really come to pass.

Some incumbents, so-called because they occupied the dominant position in their home markets when liberalization began, have lost market share. But frequently this has happened not as the result of market forces, but of government moves to cap and reduce an incumbent's market share action (as in the case of Italy's Enel) or to require an incumbent to auction off some of its gas or power to smaller competitors. Other incumbents have gained; Gaz de France already had a 90 percent share of the French gas market plus control of most storage and transmission before its government-orchestrated merger with Suez. Many have held their own. A few companies such as Germany's Eon have taken on a more pan-European character by making acquisitions in open markets like the UK and privatizations in East and Central Europe. But only in these latter markets has the concentration of dominant incumbents diminished at all.

Brussels' competition directorate reviewed the energy industry's structure in its sector study. Liberalization, it found, had
not changed the wholesale gas supply chain. 'The high level
of concentration which existed in most national markets at
liberalization largely remains.'[1] Incumbents dominate production, imports and trading on any gas hubs. But trading remains
largely localized; only two incumbents trade across Europe to
provide arbitrage and therefore price convergence.

In electricity, the level of concentration also remains high.
The early years of liberalization saw little new build – whether
by incumbents or new entrants – except for some gas-fired
plants in Italy, Spain and the UK.[2] Generators' ability to influence price comes from non-storability of electricity and highly
inelastic demand,[3] and their desire to do so stems from the fact
that the price offered to all is usually that of the most expensive.
So it can be profitable to withhold capacity if the 'loss' on the
electricity withheld is exceeded by the increase in profit on the
remaining electricity sold. Obviously there are other factors in
the setting of electricity prices – fuel costs and the impact of
emission permit trading – but the concentrated market power
of a few big generators also has a bearing on prices, feeding
distrust of the industry.

So what are the barriers to entry that have evidently shielded
incumbents from competition?

Vertical foreclosure

This is the process whereby incumbents wittingly or unwittingly
foreclose, or close in advance, the availability of crucial inputs or
assets to potential rivals. This can take several forms of locking
up energy in long term upstream supply contracts, or locking
transmission networks up in long-term capacity contracts, or
locking customers up in long term downstream distribution
contracts.

---

1   DG Competition sector inquiry, 2007, p. 37.
2   Ibid, p.134.
3   Ibid, p.132.

Table 3: How incumbents control most of the gas in their home countries

	Total imports (2004 in bcm)	Incumbents % share of imports (2004)	Total domestic production (2004 in bcm)	Incumbent % share of domestic production (2004)
Austria	9	80–90	2	–
Belgium	16	90–100	0	–
Czech Republic	9	90–100	<1	–
Denmark	0	–	10	80–90
France	49	90–100	1	–
Britain	13	20–30	105	40–50
Germany	88	90–100	18	80–90
Hungary	11	90–100	3	90–100
Italy	67	60–70	13	80–90
Netherlands	18	50–60	73	90–100
Poland	10	90–100	5	90–100
Slovakia	7	90–100	<1	–

Source: European Commission sector inquiry SEC (2006) 1724, p. 240

The issue of long term contracts (LTCs) is one that Brussels and the gas companies have argued over long and hard. Europe's gas companies, and their upstream suppliers in Russia and Algeria, regard these contracts as crucial to the planning and funding of long-distance pipelines. And the Commission does not disagree. Indeed in its latest attempt to get the gas companies to accept ownership unbundling, the Commission said, 'the key to conclude long-term supply agreements with upstream gas producers is not the ownership of the network but the existence of a strong customer basis',[4] presumably enshrined in a long term contract. But Brussels tries to insist that the LTCs do not lock up too high a percentage or for too long, though without so far really defining what that means.

The tendency or temptation for companies to use any transmission networks they may own to 'disadvantage' their rivals is what the whole unbundling saga is about. As if to dramatize

---

4   Commission explanatory memorandum of Third legislative package, September 2007.

the problem, the competition directorate launched its Eni and RWE anti-trust investigations in summer 2007. It announced an investigation into Eni's 'alleged capacity hoarding and strategic under-investment in the gas transmission system' in Italy, so shutting competitors out of some of the country's markets. A year later, in May 2008, the Commission announced a similar infringement inquiry into Gaz de France's 'behaviour that might prevent or reduce competition on downstream supply markets for natural gas in France through, in particular, a combination of long-term reservation of transport capacity and a network of import agreements, as well as through under-investment in import infrastructure capacity'.

The investigation that the Commission opened against RWE in May 2007 appears to have had a swifter, and from the Commission's viewpoint rather satisfactory, denouement. The Commission said it was investigating RWE for possibly 'abusing its dominant position' in the North Rhine Westphalian regional gas market 'by raising rivals' costs and preventing new entrants from getting access to capacity on gas transport infrastructure in Germany'. It said it suspected RWE of charging high prices for access to its gas network, inflating its network operating costs, maintaining artificial fragmentation of the network, and failing to free up pipeline capacity so that customers could switch to rival suppliers.

A year later, in May 2008, RWE said it was ready to sell its entire German gas grid of some 4,000 km, with the aim of getting Brussels anti-trust investigators off its back. The German company said its offer was not an acknowledgement of guilt. Eon had maintained its innocence in identical terms a couple of months earlier, in a similar deal with the Commission. In February 2008, Eon said it would sell off its high-voltage electricity grid in Germany in return for Brussels dropping an investigation into Eon for suspected manipulation of the wholesale and balancing markets, in which Eon's control of the grid may have played a part.

Yet many energy operators and specialists regard the vertical integration model – inherited from the old national monopolies – as a legitimate business model, especially to offset other risks in liberalized markets. In electricity, Malcolm Keay has noted that vertical integration – though not necessarily including ownership of networks – is considered a useful form of risk management. He has made a convincing case that 'the pressure to hedge risk by integrating downstream is reinforced by the special nature of electricity markets. Because it cannot be stored, an electricity sale, once missed, is lost forever. And because hedging is difficult, direct physical ownership of generating assets, combined with access to retail customers, is the easiest way of securing a match between supply and demand.' He notes that it was stand-alone 'merchant companies which suffered most in the UK and USA in the early 2000s, when many faced or suffered bankruptcy'.[5]

On a more mundane level, many ordinary energy customers may appreciate the organizational simplicity of vertical integration (even if that is not what they would call it). The argument has surfaced in an interesting way in France, whose government has been telling Brussels that all that was needed was rigorous enforcement of existing unbundling legislation (putting transmission networks into separate subsidiaries) rather than any new reform.

As part of this pitch to enforce maximum separation within the existing EU law, the French regulator, the CRE (Commission de Régulation de l'Energie), has complained about the striking similarity of the marketing logos of EdF on the one hand, and of RTE (Reseau de Transport d'Electricité), EdF's 100-percent owned but separately managed transmission subsidiary. 'Look, the network is part of our brand, whether the CRE likes it or not', says an EdF executive. 'People know the network because if anything goes wrong it is the network people that come and fix it, and customers don't want someone strange turning up looking totally unrelated with EdF when they have signed their contract with EdF. Industry may know RTE, but householders know only EdF.'[6]

In gas, the vertically integrated incumbents defend their raison d'être even more strongly, arguing that Europe needs strong players to negotiate with the world's Gazproms and to

---

5   Malcolm Keay, *The Dynamics of Power*, Oxford Institute for Energy Studies, Oxford, 2006, pp. 48–9.
6   Author interview, July 2007.

make big long term financial bets on expensive and extensive pipelines. But the Commission has accepted none of these arguments. The Directorate General for Competition claimed in its report that its public consultation on its sector inquiry findings 'has not revealed any significant synergy effects linked to vertical integration',[7] and that experience had shown that all parts of unbundled businesses continued to thrive after separation.

Sometimes, downstream contracts can damage competition as much as upstream ones because incumbents can write them to effectively lock customers in and so exclude would-be new suppliers getting a foothold. The German cartel office took a stand on this in 2005 in the case of Eon–Ruhrgas' downstream contracts. The agency had earlier tried to block Eon's takeover of Ruhrgas (see next chapter), but had been overruled by the German government. But in 2005 it got some of its own back by preventing Eon–Ruhrgas from writing new contracts covering more than four years for more than 50 percent of a customer's annual demand, or covering for more than two years more than 80 percent of a customer's demand. In 2007 the Commission reached a similar deal with Distrigaz in Belgium by getting it to agree to make some 70 percent of its gas supply to industrial and wholesale customers 'contestable' by other suppliers. Shortly afterwards, the Commission opened an investigation of Suez's Electrabel division in Belgium and EdF in France for effectively locking up their customers by making it hard for them to switch suppliers. Commission officials indicated they were looking for a 'Distrigaz' type solution.

Market segmentation

This phenomenon exists, because incumbents rarely enter other national markets as competitors. Incumbents tend to sit on existing pipeline through-put capacity via long-term, pre-liberalization capacity agreements not subject to ordinary TPA rules. And when pipelines are expanded it is generally to serve incumbents, not new entrants. In electricity, the bottlenecks

---

7   DG Competition sector inquiry, p. 14.

are less contractual and more the physical lack of sufficient interconnection between national markets. Historically, power companies only created links between each other so as to be able to help each other out in emergencies. As a goad to more action, EU leaders, at their Barcelona summit of 2002, said that each state should aim at having import or interconnector capacity amounting to 10 percent of its total generation capacity. But several countries – islands or peninsulas – do not have this: the UK, Ireland, Spain, Portugal and Italy. However, having more than a 10 percent ratio of connection that does not guarantee that countries are free from congestion or subject to competition from outside.[8] Often national politicians and regulators are chiefly or entirely to blame for this segmentation. But sometimes companies do their own segmenting.

Table 4:   Some Weak Links in Europe's Power Chain – Average hourly net import capacity (NTC) relative to electricity generating capacity 2004

Country	percent	Country	percent
UK	2	Czech Republic	23
Italy	6	Austria	24
Spain	6	Belgium	25
Ireland	6	Sweden	29
Portugal	9	Hungary	38
Poland	10	Slovakia	39
Greece	12	Denmark	50
Finland	14	Estonia	66
France	14	Slovenia	68
Germany	16	Luxembourg	90
Netherlands	17		

Source: European Commission sector inquiry SEC (2006) 1724, p.175

This is what the Commission suspected when in July 2007 it announced an inquiry into possible collusion by Eon and Gaz de France. 'The possible infringement, which will be further investigated, takes the form of a suspected agreement and/or concerted practice between Eon and Gaz de France whereby they agreed not to sell gas in each other's home market. This

8   Ibid, p. 175

agreement and/or concerted practice may concern in particular supplies of natural gas transported over the MEGAL pipeline, which is jointly owned by Eon and Gaz de France and transports gas across Southern Germany between the German–Czech and German–Austrian borders on the one side and the French–German border on the other side.'[9]

A year later, in June 2008, the Commission sent both companies a 'statement of objections', taking its investigation of their possible market sharing behaviour to a more formal stage. However, the companies still insisted that the Commission, which had conducted dawn raids on their respective premises back in 2006, had just dug up an old and out of date story. The companies claimed that the gas transport agreements that Brussels was so fussed about went back to the construction of the MEGAL pipeline in 1975, and had been terminated in 2004.

Whatever the upshot of this case, it is another instance of alleged market segmenting involving Germany. This is not surprising. Segmenting is literally built in to the German system. For it has by far the most complex energy network in terms of ownership. In electricity, it has four transmission system operators (TSOs) with ultra high voltage, 40 TSOs with high voltage, and a remaining 800 networks with medium to low voltage. Its gas network is no less of a nightmare, though the two dominant suppliers, Eon and RWE, are making a big effort to simplify it. Germany has nineteen gas balancing zones (compared to one, two or maximum three in other states). This is awkward for new entrants; the smaller the balancing zone the greater the risk of imbalance in volume and pressure that a new supplier can cause and that the new supplier has to compensate for.

The subtlest barrier, though one of the most effective, is the lack of transparent market information. New entrants find themselves at a particular disadvantage if they do not know as much, or as rapidly, as the incumbents about such vital data as network availability, infrastructure outages, or congestion in electricity interconnectors and gas transit pipelines. And in the normal course of events new entrants will not be as much 'in the know' as incumbents who, after all, control much of these

---

9   Commission press Memo/07/316, 30 July 2007.

networks and whose operations weigh most on the market. If the reason for the prices swinging around is a big utility taking a nuclear plant off line or an incumbent gas company increasing its imports, it helps to be that utility or that gas incumbent knowing this in advance. Therefore, to neutralize this advantage, there need to be means of ensuring this information is disseminated to all, and at the same time.

In terms of pursuing some of these market abuses, we have seen that the Commission has been ready to launch investigations and take, if necessary, court action. But preventing these abuses arising in the first place is not so easy. They stem from the concentrated nature of national energy markets, and that in turn is a legacy of the old national energy monopolies. What can Brussels do about this? Not much. In 2004, it did get the power to impose 'structural remedies', in plain language, to break up a company or group, 'where there is substantial risk of a lasting and repeated infringement that derives from the very structure of the undertaking'. In practice, the Commission is most unlikely to dare to use this power. Fining a company on *proof* of market abuse is something that Brussels has long done. To go on to break it up, in cold blood, so to speak, because there is a *risk* the company will keep repeating the abuse, is probably a step too far. So the Commission will continue to rely on its merger control authority for any shaping of the energy sector.

But this is a passive power. Brussels cannot use it proactively. It has to wait for companies to want to merge before intervening. Liberalization has certainly stimulated the merger market. Companies when forced to compete often choose to join rivals they can't beat. During the first phase of liberalization – from 1998 (date of the first gas directive) to 2003 (date of the Second directives) – the Commission was notified of 135 mergers and acquisitions in the electricity and gas sectors, of which two-thirds were of national dimension and one-third were cross-border.[10] In general, the Commission has taken a deeper interest in energy mergers than in other mergers. Of all mergers notified to Brussels for approval since 1994, the Commission has raised

10 Peter Cameron, *Competition in Energy Markets*, Oxford University Press, Oxford, 2007, p. 368.

queries, conducted investigations, set conditions and occasionally issued vetoes in 18 percent of energy mergers, compared to a 12 percent average for mergers in all sectors.[11]

In summary, then, it is evident that anti-competitive practices persist in Europe's energy markets. There are special economic and political features, such as the legacy of state monopolies, which have contributed to these practices. But there are also special technical and commercial features, such as synergies in vertical integrated utilities, which can mitigate these practices. Just as the issues are not always clear cut, so the remedial instruments are not always ideal. The Commission has used anti-trust investigations to the full. And it has often used its control over cross-border mergers to impose concentration-reducing measures on dominant incumbent companies such as market caps, gas release programmes or electricity auctions. But sometimes, as we shall see in the next chapter, it is governments that stitch up national mergers that are beyond the Commission's reach.

---

11 Commission officials at Casteel–Claeys seminar, January 2007.

CHAPTER 5

GOVERNMENTS BEHAVING BADLY

It is time again to stop the countries of the European Union from erecting national barriers. If not, we will risk an August 1914 effect ... with at the end, a war that no one wanted.

Giulio Tremonti, Italian finance minister, speaking on energy protectionism, February 2006.

In reacting in this way to the efforts of Paris to thwart an Italian bid for a French energy company, Giulio Tremonti appeared to be going over the top. Any such reference to the guns of August 1914 seemed quite implausible in today's Europe; for all its failings, the European Union has made war unthinkable between its members. But there was an economic relevance to his comments. For August 1914 brought to an end the world's first period of globalization without tariff barriers and currency controls, and what Mr Tremonti was really warning of was the risk of globalization going into reverse again. Nor was it implausible that the reverse might come in energy, though the greater setback was to occur in finance.

For energy was again becoming a sensitive issue for governments. In 2006, oil and gas prices were continuing their run-up since 2000. A month before Mr Tremonti spoke the supply of Russian gas had been briefly cut, in January 2006, as a by-product of Moscow's dispute with Ukraine over pricing and payment. In addition to this revived worry about energy security was the rising concern about energy as the main culprit in causing climate change. As Mr Tremonti was speaking, not only was the French government decided to push Suez into a merger with Gaz de France to keep it out of the hands of Italy's Enel, but Germany's Eon had made a counter-bid for Spain's Endesa, which the Madrid government declared itself determined to stop.

Even the UK, which had usually taken a *laissez-achêter* attitude to prospective foreign buyers of its energy companies, had fretted about Gazprom's interest in acquiring Centrica, the last sizeable UK-owned energy retailer. Newer parts of the EU energy sector have not been immune from this protectionist virus. Aided and abetted by its politicians, Hungary's MOL went to the limit of EU law, and probably beyond, to fight off a takeover bid from Austria's OMV. All this did not amount to the beggar-my-neighbour trade policies of the years after 1914, but it did constitute a sort of block-my-neighbour attitude to energy investment even from fellow EU states.

Championing national companies

Liberalization means letting go, which governments, like parents, often find hard to do. They often remain over-protective of their former energy monopolies and unwilling to surrender control over the energy sector. 'The energy sector, especially electricity, is the last tool left to governments, and they want to keep it for reasons of security of supply and environmental protection', according to Rafael Miranda, the CEO of Endesa. 'Yet in this global economy, if governments try to avoid things like unwelcome foreign bids, the strength of the market wins through in the end.'[1]

Mr Miranda should know. He was at the middle of a long corporate tug of war that started with an attempt to make Endesa part of a Spanish national champion, then nearly landed it in the German grasp of Eon, and finally put it into the hands of Italy's Enel utility and of Spain's Acciona construction company. The Endesa affair highlighted the degree to which Spain had never really accepted the consequences of its relatively early move to energy liberalization and the lengths to which it was ready to go to keep Spanish companies Spanish. It also underscored important shortcomings in EU merger controls.

The saga started in 2005, when Barcelona-based Gas Natural, prompted by Catalan nationalist politicians in alliance with

---

1   Author interview, 2007.

the socialist government in Madrid, bid for Endesa, a Madrid company considered to be more aligned with the conservative opposition. In January 2006, the Spanish competition tribunal recommended to the government that the bid should be blocked. Instead, the government overruled the tribunal and conditionally approved the Gas Natural bid. This happened just as Eon – to some extent, solicited by Endesa itself – made a considerably higher offer for Endesa. It fell to the EU merger control authorities to rule on the Eon bid by virtue of the size and trans-national character of the German company.* Brussels quickly gave its approval, because of the lack of competitive overlap between Eon and Endesa. Frantic to thwart the German bid, the Spanish government quickly passed a law to increase the national energy regulator's powers over takeover bids by foreign companies. The regulator subsequently used these powers to impose on the Eon bid a series of onerous conditions. These included obligations to use Spanish coal, to maintain the Endesa brand and to retain Endesa assets in Spanish islands and enclaves for five years.

The European Commission could not overrule Madrid's clearance of a Gas Natural–Endesa combination because it clearly came under national jurisdiction. This is a shortcoming. For national energy mergers can be clearly bad for competition, even – or one might say especially – where one company is in electricity and the other in gas. Such companies overlap, or interlock, because gas is a prime fuel for power generation and control of it tends to give an electricity company an advantage over rivals in the power business that have no gas assets.

A notorious example of such national concentration, and of Brussels' powerlessness to prevent it, came a few years earlier. Ironically it concerned Eon, which in 2001 made a takeover bid for Ruhrgas. The German cartel office blocked the bid, but was overruled by the German government, which allowed

---

* The EU vets mergers where the combined world-wide turnover is over Euros 5bn and the EU portion of this turnover is over Euros 250m for each of at least two of the companies involved, unless each of the companies does more than two-thirds of its EU turnover within one and the same state.

the bid to go ahead. And Brussels could only stand helpless on the sidelines, given its lack of jurisdiction in deals where the companies merging each do more than two-thirds of their EU business within one and the same EU state.

But Madrid overplayed its hand in the Endesa case. The Commission took the Spanish government to the European Court of Justice in two cases, one for infringing the EU merger regulation by imposing conditions on a bid by Eon that was within the sole purview of Brussels, and another for breaching EU rules on free movement of capital and freedom of establishment. In 2008 the Court upheld the Commission in both cases.

This will have sent out a wider signal of displeasure about governments shutting out companies from other EU states. For its part, the Spanish government has been chastened by its failure to keep Endesa out of foreign hands, because the company is now partly owned by Enel. Prime Minister Jose Luis Zapatero admitted ruefully in June 2008 that 'we cannot all aspire to have national energy champions.'[2] The Italian company, with its Acciona ally (a construction firm flush with cash from Spain's 20-year building boom), eventually cut a deal with Eon. It sold a quarter of Endesa (including bits in Spain and Italy that might otherwise have raised competition issues) to the German company Eon and kept the rest. So Enel got some compensation for its frustration over Suez that gave rise to Mr Tremonti's dramatic complaint and warning.

For France, there has been no such chastening experience in the construction of national champions. Despite all Paris' complaints about liberalization, its two energy giants have, on balance, gained from it. They have lost a little market share at home; according to the CRE regulator, EdF lost 6.3 percent of its customers in 2004–7 and GdF 7.4 percent over the same period. At the same time, EdF has expanded abroad with acquisitions in the UK, Germany, Belgium, Italy and Poland. The fact that EdF could expand abroad, while benefiting from a lower cost of capital stemming from its state ownership and credit rating and from the assurance of a relatively protected

Table 5:   The Endesa-Eon-Enel saga

2005

September 5	Gas Natural bids for Endesa

2006

February 21	Eon bids for Endesa
February 24	Spain gives own regulator power to stop merger
April 25	European Commission approves Eon bid
July 27	Spanish regulator imposes conditions on Eon bid
September 25	Acciona buys 10 percent of Endesa
September 26	European Commission says Spanish regulator's conditions illegal

2007

February 3	Eon raises bid for Endesa
February 27	Enel buys 10 percent of Endesa and increases this to 22 percent
March 23	Enel and Acciona announce joint bid for Endesa
March 25	Eon raises bid again for Endesa
April 2	Eon drops bid and carves Endesa up with Enel and Acciona
July 4	Spanish regulator imposes conditions on Enel/Acciona takeover of Endesa
July 5	European Commission approves Enel/Acciona acquisition of Endesa
October 19	Spanish govt modifies Spanish regulator's conditions
December 5	European Commission says modifications still breach EU law

2008

	European Commission files legal suit against Spain for not lifting modifications, and is upheld in the European Court of Justice.

home market, provoked a hostile reaction in some quarters. Italy and Spain passed what were effectively anti-EdF laws – restricting the voting rights of foreign state-owned acquirers of their companies' shares – that were eventually ruled illegal by Brussels. So, to an extent, some of the recent economic nationalism can be seen as an adverse reaction to EdF's early expansion (in the same way that Napoleon helped stir, by the act of invasion itself, political nationalism within the countries he conquered). EdF executives admit that their company gained a first mover advantage, partly by creating a protectionist reaction that has probably hampered rivals such as Eon more than EdF itself. As for GdF, Paris' defensive manoeuvre has succeeded, with President Nicolas Sarkozy presiding over a merger with Suez that is effectively a partial nationalization of Suez.

Retail price control: rigging liberalization's outcome

There is another area in which many more governments have difficulty in letting go – and this is retail energy pricing. Even as the day of full liberalization supposedly dawned on 1 July 2007 – with the opening of all retail power and gas markets to cross-border competition – several EU governments were still imposing an arbitrary lid on retail energy prices to keep them lower than market forces otherwise would.

Clearly, a few prices need regulating, but *only* those fees that owners or operators of transmission grids or pipelines charge for access to their natural *monopoly* infrastructure, and *not* those retail prices supposedly set where *competitive* supply meets demand. But many governments require their national regulatory authorities to regulate retail prices as well as access fees. In Slovakia, in 2008, a more direct approach was threatened. Its prime minister, Robert Fico, said he would renationalize the main power company, Slovenske Electrarne (now owned by Enel of Italy), if it did not curb prices increases.

The debate over energy price regulation can be dressed up in philosophical terms. In France and some other continental countries, such regulation is justified as part of the conception that energy is a special public service. This is in contrast to a more Anglo-Saxon view, which the Commission in Brussels shares, that energy is simply an ordinary commodity with a market price that should run free like any other. Yet another way to see the issue is simply in terms of political courage, or lack of it. Fearful of voters' reaction, the politicians, in the words again of Mr Miranda of Endesa, 'are not prepared to let the market work, or only when prices go down'.

The problem for the EU is not just that its 2003 directives specifically require free energy pricing, except where governments notify Brussels that regulation is necessary to fulfil public service requirements such as street lighting and to protect vulnerable communities such as the poor, unemployed or remote populations. Rather, the problem is that the EU's entire strategy needs – and is predicated on – higher prices for all conventional energy in order to stimulate energy saving and alternative energy development. Market economies need relatively high energy prices – which can be kept high by taxes whose receipts can go to the poor – in order to spur greater efficiency.

For it is the logic of Europe's emission trading system (ETS) that the cost of carbon permits should be passed on to consumers so as to discourage the use of carbon-intensive fuels. But utilities get criticised when they do pass on this cost. Only if this cost is passed through into retail prices will low carbon sources of energy, such as renewable and nuclear power, become competitive with hydrocarbons like gas or coal that otherwise benefit from a liberalized market. As Jan Horst Keppler of Paris-Dauphine university points out, 'many people are unaware of the fuel choices of liberalization which pushes companies into going for technologies, like gas and coal generation, with relatively low fixed (plant) and high variable (fuel) costs, and which disadvantages those technologies – renewables as well as nuclear – with a high ratio of fixed costs to variable costs.'[3] The only way to correct this bias of liberalization towards fossil fuels is to ensure that fossil fuel prices include the cost of the pollution they cause.

Regulating end-user tariffs has pernicious effects. To give their energy-using industries a competitive edge over others and to curry favour with consumers, many governments are setting retail prices close to or even below wholesale energy costs. Retailing power has generally been a low margin business, but regulated tariffs make it zero margin, or even a loss-maker. In several countries – Spain and France, for example – free market

3   Author interview, 2007.

prices coexist with regulated prices, and customers are free to move from regulated to free prices (though there is sometimes a restriction on moving back). In Spain and France again, companies supplying customers with power at the (generally lower) regulated tariffs get reimbursed with state subsidies for any loss in supplying the cheaper power. These subsidies tend to go to incumbent suppliers (because they are the ones with regulated-tariff customers), and not new entrants (who tend only to be able to pick up customers on the freely-priced market). This discourages new entrants. Their absence in retail markets hampers liquidity in the wholesale power markets, robbing the industry of accurate pricing signals when the need for investment in new supply is particularly pressing. Ultimately, this could force governments to start dictating investment in generation and grids, sounding the death knell for liberalization.

In launching legal proceedings in 2006 against more than half the EU membership for failing to implement the 2003 directives properly, the Commission specifically went after Spain, France, Estonia and Latvia for regulating tariffs too widely and at too far below market prices; after Italy and Ireland for granting the right to supply power at a regulated price on a discriminatory basis to incumbents; and after Germany, the Czech Republic, Poland, Slovakia, Lithuania, and Italy (only in gas) for failing to provide sufficient information on regulated tariffs.

In 2007 the Commission tried another angle of attack against two major offenders. It accused first Spain, and then (waiting until the French elections were almost over) France, of possibly violating EU state aid rules in using regulated tariffs to give their energy-using companies artificially cheap power and, as we have seen, subsidizing incumbent electricity suppliers. This weapon of a state aid investigation carries more menace, because if the governments lose the case, the corporate recipients of their largesse have to repay all of the state aid. But it is a more complicated procedure, requiring the Commission to prove that not only do the regulated tariffs constitute state aid, but that they have also distorted cross-border trade. All these procedures, too, are very long. It can take five to six years for a Commission complaint against a government to reach the stage of a European Court of Justice ruling.

Regulating the regulators

As governments have, through liberalization and privatization, reduced their direct control of European energy companies, so measures to protect the public interest and ensure fair competition, through regulation, have been developed. As one of the first to liberalize and privatize, the UK was also one of the first to set up an energy regulator, today known as Ofgem. Most other national regulators had, and still have, less power and independence than Ofgem.

At the EU level, regulation was initially little more than self-regulation. After the First liberalization package of 1996–8, the Commission set up the Florence Forum in 1998 for regular discussions with all the stakeholders in the electricity industry, and a year later the Madrid Forum to do the same for the gas industry. Lacking any law-making or regulatory enforcement powers, these forums have been essentially regulation by cooperation. The Second package of 2003 required every member state to have a specific national regulator for energy, which by that date they all did except for Germany. The Third package of 2007 called for the creation of a new semi-federal network of regulators at the EU level – an Agency for the Cooperation of Energy Regulators – as well as an upgrading and harmonization of the powers and independence of national regulators in their own countries.

EU governments have mouthed support for regulators to have stronger national and EU powers. It will be interesting to see how the politicians adjust to this in reality. The practice of farming out control of network industries like railways or communications to specialist regulators goes back to the 19th century in the USA, whose Interstate Commerce Commission (created in 1887) is the grand-daddy of them all. But it is new in Europe. Many European politicians have found it hard to swallow the idea of putting control of something as vital, and so in a way political, as energy in the hands of technocrats. Even in the UK governments tend to respond to any sharp rises in energy prices by leaning on Ofgem, the regulator, to inquire into the state of competition in the energy sector.

Naturally, the issue of regulated prices is often an irritant between government and regulator, and most acutely in countries where the tradition of independent regulation is youngest. Such tensions came to a head in Portugal in December 2006 when Jorge Vasconcelos resigned as the country's top energy regulator in protest against government intervention. In fact, Mr. Vasconcelos points out that the Portuguese energy regulator had had, from its inception in 1996, the power to fix tariffs. But, when it tried to order a price increase in 2006, the government intervened and overruled the regulatory agency by decree. 'This has created a huge tariff deficit [between electricity's generating cost and its retail price], and totally killed the free market for electricity', notes Mr Vasconcelos. 'In 2005 the free, unregulated part of the total retail power market was 25 percent, now in 2007 it is 8 percent of the total and most of the Spanish players have left.' [4] As we have seen, across the border, Spain's CNE energy regulator found itself a pawn in the government's battle to keep Eon's hands off Endesa. The latter's chief executive, Mr Miranda, complains that 'in Spain we do not have the culture of independent regulators.'

Curiously, countries' accommodation to regulation has had little to do with their degree of political centralization. For instance, France, the centralized state par excellence, has adapted quite well to regulation. Its CRE energy regulator has shown a certain independence of spirit. It has publicly complained about regulated prices, though none of its members have followed Mr Vasconcelos' example. And, during its bid in 2007–8 to ward off further structural reform from Brussels like ownership unbundling, French governments have vaunted the CRE as a model of rigorous independent regulation that the rest of Europe should follow.

By contrast, one might have thought that Germany's political decentralization would predispose it to easy acceptance of independent regulation. Not at all. Germany was the last EU state to set up an energy regulator, and what it set up, the Bundesnetzagentur, can only devote limited resources to energy because it is also responsible for the other network industries of post, telecommunications and railways. Part of German

---

4   Author interview, 2007.

regulatory reluctance in this area is that, unlike other EU states, energy has always been a private or (where publicly-owned) local affair with rules set by industry insiders, not politicians. These rules are not easy to set because the German energy industry also has far more energy networks (800 in electricity) than any other EU state. As one (non-German) executive working in Germany for a big German utility put it, 'the German system takes time to adapt, because Germans like to discuss and negotiate everything.'[5]

So, where other EU governments need to let go of their energy companies and their regulators, there is a sense in which Germany's government has needed to take charge of rationalizing its energy system. This system is important because Germany is Europe's biggest economy and main conduit for gas from the east. But Germany has constantly tried to delay any new EU reform, on the grounds that it was still digesting the last EU reform. No wonder therefore that the groans from Berlin were among the loudest when in 2007 the Commission unveiled a Third package of market reforms even before the Second package was fully in force.

---

5   Author interview, 2007.

CHAPTER 6

UNBUNDLING – UNAVOIDABLE OR UNNECESSARY?

This would be the greatest expropriation since the Bolshevik revolution.

Bruno Wallnöfer, chief executive of Tiwag, an Austrian utility.

What would people say if Heathrow were managed by British Airways?

Claude Mandil, former head of the International Energy Agency.

Of the European Commission's 2007 reform proposals, the most controversial was on ownership unbundling (OU) of energy networks. It gave vertically integrated companies a stark choice: either sell off your networks or put them under the management of separately owned 'independent system operators' (ISOs).

The potential for collective measures in Europe's internal energy market is, as discussed in earlier chapters, high, and higher than in some other federal systems, such as the US. But ownership unbundling proposals clearly pushed this potential to its political limit. Any suggestion of forced divestment was bound to raise issues of public and private property rights, and the spectre of privatization in France and of expropriation in Germany. It was, therefore, on the face of it, somewhat puzzling that the Commission should have pushed so hard. So this chapter analyses the motives and justification for the Commission's proposals, while the following chapter tracks how far the proposals got.

The Commission's Third legislative package was not, as sometimes suggested, an Anglo-Saxon plot born out of UK government pressure to spread its gospel of market liberalization. In recent years, the UK has become more interested in its continental partners' energy market structure, and in that structure conforming more to its own liberalized model. But this view was shared by all the other member states that had taken their own national decisions to unbundle the ownership of their energy networks. A dozen EU states have separated ownership of electricity networks, and seven have done so for gas networks. They included not only the northern strata of Nordic countries, the Netherlands and Belgium, but also Spain, Portugal and Italy (the latter unbundled in electricity but not gas). This entire group of countries has been concerned that by unbundling their networks, they may have put them at a competitive disadvantage with their bundled neighbours. Therefore they have strongly backed the Commission out of a concern to create a more level playing field.

Yet, even though governments tend to set more of the EU agenda than when Jacques Delors ran the Commission in the late 1980s and early 1990s, it was very much the Commission of Jose Manuel Barroso that inspired the Third package. But the Commission's energy division – DG Tren (for transport and energy) as it is known – might not have produced such radical unbundling proposals, had it not been for the involvement of DG Competition. Continental conspiracy theorists pointed to the fact that several senior officials responsible for the Third package – whether in DG Tren or the general secretariat as well as in DG Competition – were native Anglophones. Such conspiratologists, however, missed the more important point that these officials had a competition background.

Previous EU legislative proposals had been very much crafted by the energy division, which had taken a gradual approach. Over the years, as we saw in Chapter 3, the legislation sought to drive an increasing wedge between the natural monopoly of main transmission networks and the competitive sector of upstream supply and generation and of downstream distribution and sales. Integrated energy groups have successively been required to create for their networks first separate accounts, then separate managements, and eventually separate corporate subsidiaries.

But now Brussels proposed total divorce. The reason was that as a result of its competition directorate's long trawl through the energy sector, the Commission came to two conclusions. First, there was such an inherent conflict of interest in joint ownership of monopoly networks and of competitive parts of the energy business that only total separation would end it. DG Competition knew it could continue to prosecute individual companies for illegal discrimination and market abuse, and hope that such action would also frighten other companies into behaving properly. But it also realized the task was so great that only fresh legislation on ownership unbundling would suffice to create the necessary across-the-board structural change.

Second, ownership unbundling would also redress a perceived dearth of investment in cross-border networks. This is because vertically integrated networks lack the incentive, the competition inquiry found, 'to invest adequately in their networks, since the more they increase network capacity, the greater the competition that exists on their 'home market' and the lower the market price'.[1] And it was on this point that most of the economic argument about OU turned.

No one contested that OU provides, by definition, a clean and automatic end to the conflict of interest inherent in owning transmission and supply, though opponents of OU such as the French government argued that regulatory scrutiny can equally prevent bias by network owners. But the Commission was accused, from several quarters, of over-stating the causal link between unbundling and investment.

There is certainly logic to the link. Why expand your network if that just imports competition from a rival supplier? Expansion of a network can also ease the problem of bias on that network. As Joseph Kelliher, chairman of the US Federal Energy Regulatory Commission, has remarked, 'greater grid investment will make it more difficult to engage in undue discrimination and preference in transmission service, since it is more difficult to detect undue discrimination and preference when the grid itself is constrained.'[2] In other words, if a grid is congested, it is harder to tell whether the grid operator, in refusing to carry a competitor's energy across his network, is acting out of necessity or bias.

---

1 An Energy Policy for Europe, Communication from the Commission to the European Council and the European Parliament, 10.1.2007, COM (2007) 1 Final, p. 7.
2 Statement, 20 July 2006.

But is there a precise, positive correlation of the degree of network investment and of unbundling, as the Commission maintains? Is investment higher in EU states adopting OU, and lower in states where networks are still bundled?

Under-investment is certainly a concern. Given its trans-frontier remit, the Commission's main worry has been the lack of a proper energy highway system across Europe and the energy traffic jams at EU borders. The reason for this, argued Brussels, was that not enough money was being spent to expand the interconnectors which were originally designed as emergency back-up links between national energy markets, but which could not cope with the energy volumes that people were trying to trade across borders in a liberalized market. According to the Commission, 'the amounts invested in cross-border infrastructure in Europe appear dramatically low.'[3] In 2004 only 5 percent of total annual investment in electricity grids, or Euros 200m out of Euros 3.5bn, was devoted to increasing cross-border transmission capacity.

This created crowding at the borders, where demand for transmission outstripped capacity. Some of this congestion is at the borders of peripheral markets that are effectively 'energy islands' such as the UK, Ireland, Spain and Italy. These countries have an import capacity of 6 percent, or less, of their total installed generating capacity.[4] States in the Baltic region and some in southeast Europe are also similar 'energy islands'. In 2002 EU states agreed to increase their minimum interconnection levels to 10 percent (of national generating capacity). But this was only a rough rule of thumb. The Netherlands, for instance, had an interconnection ratio of 17 percent but its connectors to other markets were still almost constantly clogged.[5] The table below shows a high degree of congestion on the Netherlands' electricity links with its neighbours, as well as showing how lopsided UK power trade has been with France.

---

3   Priority Interconnection Plan, the European Commission, COM (2006) 846, p. 5.
4   Competition sector inquiry 2007, p. 175. See also Chapter 4, fig. 5.
5   Competition sector inquiry, 2007, p. 173.

Table 6: Clogged Arteries — hours with congestion as a percentage of all hours (selection of borders)

Border	2004 Jan–May	2005 Jan–May
SK→HU	100.0	100.0
FR→CH	100.0	100.0
DE→DK	99.3	100.0
NL→BE	96.4	100.0
FR→UK	94.6	95.6
DE→NL (1)	87.9	90.1
FR→ES	34.6	81.1
CZ→DE	69.2	68.0
NL→DE (1)	62.9	63.9
BE→NL	63.3	63.1
DE→FR (1)	0.0	41.3
CZ→AT	0.0	37.0
DE→CZ (1)	30.0	35.7
UK→FR	31.5	35.0
FR→DE	48.4	33.3
ES→FR (1)	30.0	32.8
PL→SK	0.0	19.1
ES→PR	7.8	17.5
PL→CZ	15.8	16.1
PR→ES	26.7	11.7
FR→BE	30.4	11.0
CZ→PL	0.2	10.1
SK→CZ	1.4	6.6
CZ→SK	2.1	1.1
DE→CH (1)	0.0	1.0
FR→IT	0.7	0.8
AT→CZ	0.0	0.3
CH→FR	0.0	0.0
IT→FR	0.0	0.0
BE→FR	0.0	0.0
DE→AT	0.0	0.0

Source: European Commission sector inquiry SEC (2006) 1724, p.173

Note: Hours when requested capacity exceeded available cross border capacity as a percentage of all hours. The arrows indicate the direction per border, in some cases reported by different TSOs. (1) Refers to an average of more than one interconnector between two adjacent borders.

The requirement for much more network investment in the future is also clear; the Commission has forecast a need for a minimum of Euros 30bn by 2013 (Euros 6bn for electricity, Euros 5bn for LNG terminals and Euros 19bn for gas pipelines).[6] Better infrastructure to link up Europe's national markets will be needed, among other things, to improve security by allowing gas to be moved around in a supply emergency; to encourage cross-border trading, competition and eventual price convergence; and to get pan-European economies of scale in developing and trading renewable energy at least cost.

Investment

Yet what was the evidence that ownership unbundling (OU), where EU states have adopted it at the national level, has increased investment? The Commission saw 'a significant and constant increase', with an actual doubling of investment spending in the case of Spain, the Czech Republic, Portugal and the Netherlands for both gas and electricity.[7] Investment figures on bundled networks in France, Germany and Italy (gas alone) are – perhaps for obvious reasons – harder to find. They showed an increase, though less marked than in OU states. Ergeg, the European Regulators Group for Electricity and Gas which broadly supports OU, cited the interesting example of Portugal, showing a fall and then a rise in power transmission investment from 1994 to 2006.[8] This period spanned one year of vertical integration, five years of legal unbundling (network put into separate subsidiary) and seven years of full ownership unbundling.

The Commission did not make much of UK figures to bolster its case, but one of its critics, Professor Philip Wright of Sheffield University, did in order to argue the opposite. He contended that too many extraneous factors (such as planning, economic cycles, supply gaps, fuel costs) go into decisions on

6   Priority Interconnection Plan, the European Commission, COM (2006) 846, p. 5.
7   Commission impact assessment, SEC (2007) 1179, Annex 111, p. 90, 10.1.2007.
8   Ergeg, Report on Unbundling, June 2007, p. 37, fig. 3.

Table 7: Unbundling and Investment in transmission system operators (TSOs)

TSO	Member State	Currency	1995	1996	1997	1998	1999	2000	2001	2002	2003	2004	2005	2006
Red Electrica de Espania	ES	Investment in transmission grid (mill. Euros)								203	215	243	420	510
Terna SpA	IT	Investment in tangible fixed assests (mill. Euros)							191	164	240	278	259	319
CEPS	CZ	Investment (excl. financial invest.) (in mill. CSK)						628,3	781,2	506,6	1388,3	1232,2	1462,8	2348,1
Lietuvos Energija AB (electricity)	LT	Investment (mill.LTL)								120	149	145	129	156
National Grid Electricity Transmission plc	UK	Replacement, reinforcement, growth and extension (mill. £)							371	391	426	395	526	
Tramission plc Gasunie incl. GTS	NL	Investment in grid & storage (not LNG (mill. Euros)			104	84	70	57	67	83	97	114	257	529
Enagas	ES	Investment in grid, LNG and storage (mill. Euros)								192	426	463	359	433

Table 7: *continued*

TSO	Member State		Currency	1995	1996	1997	1998	1999	2000	2001	2002	2003	2004	2005	2006
Transco/National Grid Gas	UK	Investment	(mill. £)			147	191	140	228	239	182	159	128	359	444
REN	PT	Investment	(mill. Euros)	93,7	90,6	76,4	57,1	63,9	54,9	81,7	110,3	127,1	144,4	222,2	243,7
GDF/GRT	FR	Network investment in France	(mill. Euros)									970	983	1200	1400
Snam Rete Gas	IT	Investment	(mill. Euros)							429	385	505	574	685	675
RTE	FR	Investment	(mill. Euros)							651	616	535	538	582	638
All German Electricity TSOs	DE	Investment in network only	(mill. Euros)	3600	3100	3000	2700	2500	2000	2200	1800	1700	2000	2000	2500

Explanation: shaded cells indicate years in which the respective companies were ownership unbundled.

Source: Commission impact assessment, 2007, SEC (2007) 1179, page 90

new investment for new investment to be a reliable measure of investment effort. Focusing on gas, he suggested stripping out extra import infrastructure spending on the grounds that it would have been built whatever the policy regime (simply because the UK has an increasing trade deficit in gas). Instead, he used replacement expenditure as the steadiest, and therefore best, measure of investment effort. This showed that in real terms, UK replacement spending on its gas network declined from nearly £600m a year in the mid-1980s, when the network was still part of a bundled state utility, to half that level in the 2000s when it had become an unbundled private company.[9]

Yet while Philip Wright may be correct that the quicker than expected decline in the UK's North Sea gas fields would have created a distorting acceleration in investment, it is hard to see how replacement investment is necessarily a much steadier measure. Surely, there must also be a cycle in replacement expenditure, in line with the ageing of equipment. However, Mr Wright provided a useful reminder of the many other factors going into new investment decisions, some may be at least as important as unbundling. One obviously important factor is the degree of social and environmental acceptability of new power lines or gas pipes; it now takes seven to ten years to erect any line of high voltage pylons longer than 10km, or considerably longer than building, for instance, a new gas-fired power plant. Another is the attitude of regulators. This could even be decisive because it is national regulators who have to approve investment plans and set the rate of financial return.

Congestion revenues

The Commission argued that a clear indicator of the link between OU and investment was to be found in relative use of congestion auction revenue. In other words, what the various transmission operators did with the proceeds of the scarce capacity they auction off. A survey conducted by the Commission

---

9   Professor Philip Wright, Presentation to Eurogas conference, Brussels, September 21 2007.

covering the EU-15 (excluding the ten new member states) between 2001 and mid-2005 found that OU states reinvested 33 percent of congestion revenue to try to remove the bottlenecks giving rise to that revenue, compared to only 16   percent for bundled states.

Table 8:   Relationship between Ownership of TSOs and Reinvested Congestion Revenue

	Ownership unbundled TSOs in EU-15	*Vertically integrated TSOs in EU-15*
Congestion revenue (2001--6/2005)	387	623
Interconnector investment	129	104
Share of reinvested congestion revenue	33.3%	16.8%

Source: Commission impact assessment, 2007, SEC (2007) 1179, p. 34.

Focusing on Germany, the Commission found that over the same period three of the country's four vertically integrated TSOs generated Euros 400–500m in congestion revenue, but only reinvested Euros 20–30m of this into reinforcing and expanding the grid and interconnectors.[10] The TSOs apparently spent the rest of the money lowering transmission tariffs. There was nothing wrong with that, except that it did nothing to remove the bottleneck. However, it is important to recall the same caveat about regulatory, social and environmental obstacles to building new transmission. So reinvestment of much of the congestion auction money might have been pointless, because virtually impossible.

LNG terminals

The Commission used the disproportionate number of terminals built or planned in two ownership unbundled states, Spain and

10 Commission impact assessment, SEC (2007) 1179, p. 34.

the UK, in support of its general argument. But this evidence was too slender, and might just reflect that both countries face a 'gas gap' and have the ports to easily accommodate LNG ships. Italy had an even bigger 'gas gap', and Eni, the vertically integrated, dominant incumbent of its still-bundled gas industry, has only built one LNG terminal. This would seem to bear the Commission argument out. Yet there have been many other efforts to build LNG terminals in Italy that have foundered on purely environmental objections.

Market concentration and prices

The Commission made a connection between OU and lower prices. It did so by arguing that OU had the effect of weakening the market power of vertically integrated incumbents because removal of their vertical integration encouraged new entry, competition and thereby lower prices. And it cited some striking figures for its case.

One set of figures[11] charted the evolving market share of the largest electricity company in a variety of OU and non-OU states over the 1999–2005 period. The average share of the biggest generator in non-OU states was 73 percent in 2005, compared to only 47.7 percent in OU states. However, as the Commission itself admitted, this difference largely existed before any countries adopted OU. In addition, some states such as Italy, which has implemented OU in electricity but not yet in gas, accompanied the reform with an explicit order to its dominant incumbent, Enel, to reduce market share.

Moreover, the main German power producers counter-attacked against the Commission's complaint that they are an oligopoly with significant market power. In particular, the RWE utility commissioned one of Germany's better known economists, Axel Ochenfels, to review the study of the German wholesale power markets which the London Economics consultancy had carried out on behalf of DG Competition in Brussels. Professor Ochenfels complained, among other things,

---

11 Ibid, Annexe 11, p. 89, fn. 7.

that London Economics had failed to take account of the competitive effect of cross-border trading, which was 'particularly applicable to the German electricity market as a consequence of its central position', and ignored the fact 'prices and price movements in Germany, France and Austria are at times virtually identical despite major differences in cost structure.' This', said Mr Ochenfels, 'indicates that the market is to a large extent integrated'.[12]

Table 9: Development of Market Shares after Unbundling – Market share of the largest generator in the electricity market

	1999	2000	2001	2002	2003	2004	2005
Countries with legal unbundling							
Belgium	92.3	91.1	92.6	93.4	92	87.7	85
Estonia	93	91	90	91	93	93	92
Ireland	97	97	96.6	88	85	83	71
Greece	98	97	98	100	100	97	97
France	93.8	90.2	90	90	89.5	90.2	89.1
Latvia	96.5	95.8	95	92.4	91	91.1	92.7
Hungary	38.9	41.3	39.5	39.7	32.3	35.4	38.7
Poland	20.8	19.5	19.8	19.5	19.2	18.5	18.5
Average	78.8	77.9	77.7	76.8	75.3	74.5	73.0
Germany (largest)	28.1	34	29	28	32	28.4	n/a
Germany top 3			63	66	66	66	66
Germany top 5			72	75	80	80	79
Countries with ownership unbundling							
Czech Republic	71	69.2	69.9	70.9	73.2	73.1	72
Denmark	40	36	36	32	41	36	33
Finland	26	23.3	23	24	27	26	23
Italy	71.1	46.7	45	45	46.3	43.4	38.6
Lithuania	73.7	72.8	77.1	80.2	79.7	78.6	70.3
Portugal	57.8	58.5	61.5	61.5	61.5	55.8	53.9
Slovakia	83.6	85.1	84.5	84.5	83.6	83.7	83.6
Spain	51.8	42.4	43.8	41.2	39.1	36	35
Sweden	52.8	49.5	48.5	49	46	47	47
United Kingdom	21	20.6	22.9	21	21.6	20.1	20.5
Average	54.9	50.4	51.2	50.9	51.9	50	47.7

Source: Commission Impact Assessment, 2007, SEC (2007) 1179, p. 89

As regards prices, the Commission measured electricity price trends across the Union's entire 27-country membership. It found that over the 1998–2006 period, power prices to industry fell by 3 percent in OU markets, but rose 6 percent in non-OU countries. The difference was greater in electricity prices for households, which rose by 5.9 percent in OU countries and by 29.5 percent in non-OU markets.[13] This last measure is, however, unreliable. The practice of governments regulating electricity prices for households is widespread (see Chapter 5), and cuts across the OU divide within the EU. Spain was, for instance, an early introducer of OU, but heavily regulates domestic power prices. Nor in absolute price terms is there a clear difference between OU and non-OU states. Italy charges the highest prices in Europe for its unbundled electricity, because it has phased out nuclear power, makes no use of what coal it has, and is chronically short of gas for power generation. Meanwhile, France, the leading opponent of OU, has had below average power prices, thanks to its nuclear programme.

In sum, there is probably a connection between OU and lower prices, but a fairly weak one. There are so many other structural factors affecting prices such as a country's energy mix, and cyclical factors such as variations in the oil price to which most continental gas prices are pegged. And the link could be cultural as much as anything else, in the sense that the sort of country that cares enough about competition to introduce OU is likely to be the sort of country that tends to wants to see consumers' interests put ahead of producers' interests in the form of lower prices.

OU and damage limitation

In addition to vaunting the merits of OU, the Commission were keen to minimize any negative effect that sale of networks

12 Axel Ochenfels, 'Measuring market power on the German electricity market in theory and practice – critical notes on the LE study', *Energiewirtschaftliche Tagesfragen* 2007, 52(9), pp. 12–29.
13 Commission impact assessment, SEC (2007) Annexe VII, p. 105.

Table 10:   Unbundling and Electricity Prices – Electricity price evolution in all 27 EU Member states

| | *Industry* (Annual consumption: 30 MWh) | | *Households* (Annual consumption: 600 kWh) | |
	MS with ownership unbundling	*MS with integrated TSOs*	*MS with ownership unbundling*	*MS with integrated TSOs*
1998 1st half	100.00	100.00	100.00	100.00
1998 2nd half	97.38	98.88	99.75	99.68
1999 1st half	92.88	99.25	93.72	101.51
1999 2nd half	92.18	97.60	92.17	101.57
2000 1st half	88.77	97.59	94.98	102.02
2000 2nd half	88.71	94.56	93.77	101.73
2001 1st half	85.69	93.92	92.59	103.00
2001 2nd half	85.59	95.79	95.05	101.78
2002 1st half	74.76	96.28	97.54	110.12
2002 2nd half	73.01	93.60	93.52	109.66
2003 1st half	79.86	99.36	93.91	114.80
2003 2nd half	73.82	101.59	93.97	112.70
2004 1st half	81.00	101.85	94.32	113.84
2004 2nd half	79.25	106.23	92.64	115.55
2005 1st half	83.78	104.86	96.10	122.31
2005 2nd half	85.47	106.41	96.28	124.49
2006 1st half	93.63	111.92	103.23	128.29
2006 2nd half	96.99	106.01	105.91	129.46

Source: Commission impact assessment, 2007, SEC (2007)1179, p.105.

might have on the value of the previously vertically integrated group. Here the Commission's evidence carried reasonable conviction.

Some opponents of OU, especially in Germany, likened it to expropriation. Yet the Commission analysis was that 'shareholders have in fact in almost all cases benefited from increasing share prices during and after the ownership restructuring.'[14] Another fear was that without the secure and steady income stream from their network businesses, the credit ratings of

14 Ibid., p. 35

(previously) vertically integrated groups would take a hammering. In practice, there was little to choose between the credit ratings of OU and non-OU companies.

The fear of losing network ownership was particularly acute in the gas sector. The European gas business is increasingly an import business. So pipelines account for an ever larger part of the assets of gas companies. Many of the latter therefore feel very nervous about the OU plan, but not all perhaps.

In this context, it is worth bearing in mind certain advantages that some OU companies or indeed countries have had over others. It was, for instance, undoubtedly easier to unbundle the gas grids of countries that were themselves gas producers such as the Netherlands and the UK, because such gas grids would not have had to worry about their commercial weight in relation to foreign suppliers. It must also be easier for a gas company to contemplate losing its network if it is itself part of a much larger whole involved in the more valuable petroleum sector. As examples of the latter, consider Spain's Gas Natural which is part owned by Repsol, or the controlling stake that Eni has in the Snam Rete Gas network that is only a very small part of the Italian oil major's market capitalization.

In fact, concerns about the relationship between unbundling and corporate size surfaced towards the end of the controversy among EU governments. Some member states feared that ownership unbundling would leave their smaller stand-alone networks or supply businesses as easy takeover prey for the superior financial firepower of larger, vertically integrated giants based in states that still allowed bundling. Memory of Electricité de France's foreign acquisition spree a few years earlier lent substance to this fear. So Dutch and Spanish ministers successfully insisted in the negotiations of 2008 that ownership unbundled states should be allowed to take safeguard measures – which would have to conform to EU single market rules and be justified on energy supply grounds – to ensure a level playing field. The UK indicated that it would not be resorting to such safeguards because it remained relaxed about foreign companies buying almost whatever of the UK energy supply sector they wanted, provided they paid a good price for it.

ISO: the unpalatable alternative

The Commission had clearly designed the option of an 'independent system operator' to be as unattractive as possible, so that as few countries as possible would choose it instead of OU. In this the Commission succeeded so well that no one else could stomach the ISO option, not even the regulators.

The Commission wanted ISOs to be as 'deep' as possible. These operators were to have a decisive say in the development and expansion of networks that would leave the actual owners of the grids or pipelines as financial holding companies with no day to day operational control over their assets. The Ergeg regulators' group saw a very awkward division between network operators and owners in sharing legal liability for any failure to deliver or damage caused, in sharing profits, and in agreeing investment. 'Effectively this would imply the regulators' deep involvement in the investment planning and approval process', Ergeg said warily.[15]

The regulators group cited a couple of country case studies to rub in the inferiority of ISOs to OU:[16]

- Italy created an ISO in electricity in 1999. But after several years of inefficient and difficult coordination between the owners and operator of the grid, Italy 're-bundled' grid ownership and operation together again in 2005 and spun it off as an independently-owned network company. This new network company was able to reduce spending on operations and raise it on investment.

- The UK is unique in the EU in harbouring both systems of OU and ISO. In England and Wales, National Grid/Transco owns and operates the wires and pipes, and also acts as the ISO of Scotland's electricity grid which the two Scottish electricity companies still own but no longer run. Ergeg, or its UK member, Ofgem, complained the complicated interface between operator and owners in the Scottish system took 12 months and 200 pages to codify, and that there were still inefficiencies in the Scottish system.

---

15 Ergeg, Report on Unbundling, 5 June 2007, p. 14
16 Ibid, Annexe 3, pp. 32–40.

Lessons from the US

Europe is not alone in struggling with the twin problems of discrimination and under-investment on networks, particularly electricity. Indeed, in the US the problem may be worse. According to the US Federal Energy Regulatory Commission (Ferc), transmission investment declined in real terms for 23 straight years, from 1975 to 1998; it has increased since 1998 but it is still below the 1975 level.

The Ferc has never tried to order utilities to sell off electricity grid assets. It would never dare to try to do that, in the face of entrenched states rights and for lack of the federal authority that it has in the natural gas sector. Instead, the national US regulator tried something very similar to what the European electricity industry was to propose in 2007 as an alternative to ownership unbundling. In 1999 the Ferc decided to start encouraging the voluntary banding together of companies owning network assets in 'regional transmission organizations' (RTOs) that would be regulated by the Ferc itself. When this did not work, the Ferc responded with its 'standard market design' proposal of 2002, making it mandatory for companies to form RTOs. This drew a wave of opposition from an industry that is even more diverse in size and nature than Europe's, with ten times the number of transmission companies. The loudest outcry came from the southeast's vertically integrated giants of Southern and Entergy (America's equivalent of EdF) and the publicly owned companies of the Pacific northwest, but the fear among states that RTOs would allow neighbours to 'steal our power' was widespread.

So the Ferc abandoned the attempt to compel companies into RTOs, though much of the country where the congestion was worst (the northeast, Midwest, part of California and Texas) has adopted this model. But even for some former Ferc commissioners such as Nora Brownell, the creation of RTOs without separating ownership of the grid and generation has been a second best. Ms Brownell, like Brussels, believes that there are pitfalls in such wider regional organizations. "They can work efficiently, but they also need to be watched for collusion." [17]

---

17   Author interview, February 2007.

In fact, US-style RTOs do not quite fit the category of what the Commission defined as its 'second best' solution – independent system operators with the clout to order the expansion, not just the operation and maintenance, of a grid, even though it does not own the grid. "None of the US RTOs has the authority to order investment, though they can through a convoluted process divide projects into those justified on 'reliability' and on 'economic' grounds", says Ms Brownell. The idea is that 'reliability' projects (necessary to keep the grid running) get priority over 'economic' ones. "But often anything the dominant utility wants is classed as 'reliability', and everyone else's projects get downgraded as 'economic'", comments Ms Brownell.

Another reason for European caution in looking at America's RTOs is that most RTOs go beyond what their European counterparts would do to operate a market as well as a grid. Indeed some of them, like the much-vaunted PJM with its 50m customers in the mid-Atlantic states, started as power pools. There are advantages to running a market as well as a grid, says Branko Terzic, another former Ferc commissioner, "in the form of integrated transmission and generation rules and market knowledge that allow you to pinpoint where new transmission is most needed".[18] The US model may only be relevant if European transmission companies also turn themselves into wholesale marketers.

The industry's alternative

However, this regional US model found an echo in the response of the main organizations of European energy producers. As the power sector's main association and one dominated by historic incumbent producers, Eurelectric initially came out against the Commission's structural reforms. But it quickly responded with a regional recipe that, whatever its technical merits, seemed politically astute. The association told the Commission that its approach was too national, an accusation calculated to hit home with Europe's civil servants. Eurelectric said ownership

---

18 Author interview, February 2007.

unbundling 'would not of itself lead to the development of competition on a larger scale and would, moreover, reinforce the prevailing excessive national focus as identified by the Commission in its analysis', and 'is shifting attention away from the core requirements for fostering market integration'.[19]

Effectively, Eurelectric was telling the Commission that, if its top priority was market integration, it should tackle this directly by encouraging transmission systems, whatever their ownership, to cooperate across borders, and quit its quest to push unbundling to a counterproductive extreme. There appeared to be a beguiling logic to this, which was backed up by convincing lobbying and concrete example.

Particularly interesting was the attitude of Vattenfall. This company owned no transmission in its home of ownership-unbundled Sweden, but did own a network in Germany where it has been one of the dominant four power companies. 'We are ready to commit our TSO [in Germany, for instance] to a regional, i.e. supra- national, transmission structure', Vattenfall wrote in a formal company position paper in February 2007, 'but we will not support our TSO becoming part of a national ownership unbundling or ISO solution.' The Swedish company said its experience in the Nordic market was that 'ownership unbundled TSOs are driven purely by national regulation encouraging them to move congestion to the national border, instead of making the investments needed to integrate the Nordic market.'

Opposition to OU by some other Eurelectric members, such as leading French and German companies, was more predictable, because it was rooted in the culture of their home countries and governments. However, French and German governments, regulators and companies showed they were ready to take a more direct route to market integration. In June 2007, they joined their Belgian, Dutch and Luxembourg counterparts in committing themselves to an eventual coupling of their five national power markets (there is already a trilateral coupling of the French, Dutch and Belgian markets). Although more a matter

---

19 Eurelectric position paper on the European Commission's approach to market developments, March 2007.

of organization and coordination than of new investment, such market coupling is a very important market development on the ground. For it allows buyers and sellers to 'couple' instantaneously and automatically the cross-border purchase of power with the acquisition of physical actual capacity to actually get the power across the frontier. This avoids the not infrequent situation in which a cross-border transaction falls through purely for lack of immediately available transit capacity.

Yet the Commission has remained generally unswayed by such arguments or apparent bona fides. Brussels tended to claim that letting unbundled TSOs band together regionally would be at best a diversion from OU and at worst the opportunity for collusive market sharing and rigging.

Conclusion

The balance of economic cost/benefit balance pointed in favour of OU, but not as decisively as the Commission maintained.

It is logical to imagine that bundled networks might have an incentive to under-invest, though the evidence is slim. The one classic case has been the Italian anti-trust authority's decision in February 2006 to fine Eni Euros 290m for stopping upgrades on the Trans Tunisian Pipeline, even though the Italian company had already signed up a number of gas shippers to transport their gas through the expanded pipeline. The accusation was that Eni decided, on second thoughts, it did not want the pipeline to import more competition to Eni's own gas supply operations in Italy.

Again, it is logical to think that OU networks would, if anything, try to over-invest in transmission because that is their only business. Actually, over-investment could be almost as much against consumers' interest as under-investment, if the cost of unnecessary spending were passed on to consumers. A regulated utility can find it profitable to 'over-invest' provided the regulator allows it to recoup the cost of the over-investment from consumers. Moreover, it is also possible to imagine regulators being more lenient on unbundled networks over-investing than on bundled networks under-investing.

Ergeg made clear it preferred the simplicity of OU. It is easy to see why regulators might want to avoid being dragged into organizing the complexity of ISOs. The question therefore arises whether regulators are simply rewarding the system they are most comfortable with. Such a possibility is reinforced by the European Commission's comment, in its impact assessment of investment, that: 'the independence of ownership unbundled TSOs from supply and generation interests is likely to have contributed to the regulators' willingness to finance the investments through tariffs.'[20]

There is no hard proof of this. One UK regulatory official detects a cultural difference in the approach to bundled and unbundled networks. The attitude to the former tended to be 'cynical', with the regulator always wondering whether revenue allowed the network would not actually disappear into some black hole in the parent company; by contrast, the regulator can afford to be more 'trusting' towards an OU network. But while regulators have it in their power – through setting the levels and rates of return on investment – to reward whichever structure suits them best, there is no actual evidence they do.

In the end, the Commission's competition inquiry report into the energy sector and its impact assessment of the Third package provided intellectual ammunition for the OU cause, but not the knock-out argument Brussels officials had hoped for. So it was not surprising to see, as the unbundling debate reached its political as well as intellectual limit, the Commission's initially bold legislative assault slow and end in compromise, as we shall see in the next chapter.

---

20 Commission impact assessment, SEC (2007), p. 34.

CHAPTER 7

CONFRONTATION AND COMPROMISE

The coincidence of these events, at a moment when the Commission is trying to force through a very sharp position against a minority, is a very questionable game...

Peter Hinze, deputy German economy minister, on learning the Commission had pressured Germany's Eon into selling its power grid just as he walks into an EU energy council to oppose ownership unbundling.

The potential for collective restructuring of Europe's energy market may be high. But, with the concept of energy market liberalization already more than 20 years old in Europe, there were unlikely to be many more member states willing to bow to an EU directive to adopt full ownership unbundling (OU), if they had not already done so of their national volition. So the Commission knew it could not take a rigid template approach.

The EU executive's usual role is to act as a good shepherd in trying to keep its flock of member states together, though the exercise is more like herding cats than sheep. Often, in the EU legislative process, opt-outs and opt-ins are created to satisfy one or other recalcitrant states, partial forms of closer integration develop over time such as in monetary union or judicial cooperation, and some states, usually the less economically advanced ones, are granted temporary exemptions from new EU laws in the form of derogations. Sometimes, too, individual member states so exploit the degree of national discretion and flexibility allowed them in a directive that the legislation ends up as more loophole than law; a case in point is the EU directive on corporate takeovers.

But what was unusual was the Commission's decision, from the outset, to offer governments a choice between separating the ownership or the operation of gas and power networks from other parts of the energy business. This, too, could be a permanent choice. OU was clearly considered the default mode that any rational government should want to choose, and the 'independent system operator' (ISO) was clearly considered the second best option that came with additional red tape and was referred to as a derogation from the OU norm. But there was no indication that the ISO derogation would be limited in time.

So the Commission recognized the need early on to cater for two irreconcilable camps (on OU). Its January 2007 energy proposals contained the two options, which then reappeared in the draft legislation formally unveiled in September 2007. Not a lot changed over that nine-month period. The two camps could agree on the need for 'effective' unbundling without defining what effective might entail. This was the vague formula endorsed by energy ministers meeting in February, and repeated by heads of government at their March 2007 summit in Brussels. OU enthusiasts in the Commission chose to declare that the March summit had basically backed the only option that they themselves judged effective. In fact, the March summit really did not advance the debate one way or the other, because it devoted all its time and energy to issues of climate change and renewables (see Chapters 10–12).

During the first part of 2007, the electricity industry did make a serious effort to deflect the Commission away from its proposals. Superficially, it was surprising that this effort should have been made by the electricity sector rather than the gas industry, for the latter had always been more rooted in its opposition to unbundling its networks. But partly because of this, the gas sector tended to take a more head-in-the-sand view of the Commission's proposals, hoping that somehow they would just go away. The Eurelectric industry association, by contrast, was more proactive. It came up with idea of the regional independent operators (RIOs), claiming that this would kill two birds with one stone, by achieving independence and integration simultaneously. It was a clever approach, because it seemed to tackle the integration issue more directly than unbundling. But initially, Commission officials argued that RIOs would make the situation worse. If energy groups were allowed to retain their vertical integration and band together in RIOs, Brussels officials said, they would have a mechanism for collusion.[1]

In public at least, there was only one serious wobble on the part of the Commission, or rather from Andris Piebalgs, the energy commissioner whose departmental officials had, mostly, never been as keen on the OU option as their counterparts in DG Competition. The wobble came after the June 2007 energy council, the first and only time that Germany had allowed serious discussion of energy reform during its presidency of the EU in the first half of 2007.

To widespread surprise, even among his own officials, Mr Piebalgs told a press conference after the meeting that 'I can clearly conclude the majority [of member states] is not with me', and that this 'very uneasy situation...has become much more serious for me.' Yet officials from states on both sides of the ownership unbundling all agreed that Mr Piebalgs was over-reacting to his opponents. The latter had not grown in number. Indeed there was one switch in that evenly-split energy council meeting as Poland moved from the negative to the positive camp.

Mr Piebalgs may have been daunted by the more formidable display of Franco-German opposition to ownership unbundling on that day in June 2007. France fielded a former prime minister Alain Juppé, who was briefly its energy minister. Meanwhile Germany departed from the studied neutrality it had shown earlier during its six-month presidency of the EU. German economics minister Michael Glos first encouraged his deputy Joachim Würmeling to voice Germany's national opposition to unbundling and then, in his own supposedly balanced summing up as council chairman, was himself dismissive of ownership unbundling.

But Mr Piebalgs' public doom and gloom galvanized his ministerial supporters. A fortnight later, ministers from eight pro-OU states wrote him a sort of 'anti-depressant letter', sent by Danish energy minister Flemming Hansen and co-signed by his counterparts from Belgium, Finland, Netherlands, Romania, Spain, Sweden and the UK.[2] It told Mr Piebalgs and fellow

1   Commission press conference, 19 September 2007.
2   Author copy of letter from the Danish energy minister and others to Piebalgs, 22 June 2007.

competition commissioner Neelie Kroes that the eight ministers believed the June 6 council displayed 'a clear majority in favour of ownership unbundling at transmission level'. This last phrase was a reference to the fact that several countries which had been sitting on the fence, such as Poland, had said they could support ownership separation provided, as the Commission had promised, it only applied to main transmission networks and not to smaller final distribution systems.

OU would 'ensure the best possible incentives for investments in infrastructure and non-discriminatory behaviour', the eight ministers wrote. No alternatives had been proposed that would 'eliminate built-in conflicts of interest' and 'avoid overly detailed and complex regulation', they added.

On receipt of the letter, a spokesman for Mr Piebalgs said the Commission was pleased to see 'substantial and outspoken support' for OU. Its opponents, however, did not take this lying down. At the end of July Mr Piebalgs got a letter from France, Germany and seven smaller countries (that included Mr Piebalgs' own Latvia, as well as Austria, Bulgaria, Cyprus, Greece, Luxembourg and Slovakia). Many of these countries were worried about the suitability of OU in small or isolated markets. The letter questioned how OU would help lower prices or raise investment, and said 'the idea of complete separation of production and transmission as the only key to the development of the internal energy market for electricity and gas should be avoided.'[3]

However, this apparent impasse in the Council was not paralleled in the European Parliament, which under the co-decision procedure has an equal say in legislating on single market matters. Most MEPs leaned towards unbundling, as shown in a vote in July 2007 on a non-binding resolution approving a pro-ownership unbundling report by Spanish conservative Alejo Vidal-Quadras. In spite of the fact most MEPs usually follow the voting line decided by their trans-national political grouping, on unbundling a number sided with their governments. One of those to take a line very different from his government was Claude Turmes, a Luxembourg Green, who like the rest of

---

3   Author copy of a letter from the nine energy ministers to Piebalgs, 27 July 2007.

his group, suspected the big utilities of discriminating against renewable energy on their networks.   After the vote on the Vidal-Quadras report, Mr Turmes crowed: 'The mobilization of French and German [MEP] colleagues by energy giants EdF, Eon and RWE to defend their own financial interests has failed.'

These, then, were some of the political forces that shaped the package of draft legislation that the Commission unveiled on September 19 2007. This had the following main elements:

- Unbundling: emboldened by the support of national energy regulators, more than half the MEPs and at least half Council members, the Commission stuck to its preferred prescription of ownership separation, but with an opt-out for states to let their energy groups retain title to their networks but cede their day to day control to 'independent system operators'. Eurelectric's proposal of regional independent operators figured nowhere.
- Safeguards against unwelcome foreign investment. The Commission proposed a two-step safeguard against companies from third countries seeking to acquire a significant stake or control over EU energy networks. Such third countries would have to have an agreement with Brussels explicitly allowing this kind of investment by their companies in the EU. Even then, national regulators and the Commission could still review, and possibly block, their investment. This was designed to take care of the growing concern about Gazprom.
- Regional solidarity agreements by groups of EU states to plan sharing of stocks in the event of energy supply cut-offs. Pressure from East European states for better energy security arrangements thus appeared to have partially paid off.
- A series of measures to promote cross-border trade and investment. These included strengthening and guaranteeing the independence of national regulators who would also get a new coordinating body (the Agency for the Cooperation of Energy Regulators or Acer). Transmission system operators would also get a new body to plan new investment.
- Greater transparency in the flow of information relating to network operation and supply.

Predictably, and despite the real importance of the regulatory and transparency reforms, attention focussed on the unbundling proposals and particularly on the Commission's impact assessment justifying its proposals. The Commission's opponents, notably France and Germany, set to work to criticise the impact assessment, and with some effect. They briefed MEPs, particularly French and German ones, on the perceived shortcomings of the impact assessment (discussed in detail in the last chapter), and in an autumn hearing the Parliament's energy committee gave Commission officials a rougher ride than they had been used to on the unbundling issue.

If the Commission had hoped that the hard line opponents of OU such as France, Germany and Austria would grasp at the ISO option, it was quickly disabused. For while the opponents might regard the ISO option as the lesser evil (because it did not involve forced sale of network assets), some of them regarded it as an even bigger nonsense, reducing a network owner to nothing more than a financial holding company.

Initially, France and Germany asked the Commission to come up with 'a third way'. Finding the EU executive unwilling to do so, Paris and Berlin then decided to work out their own alternative, together with the seven countries that joined them in signing the July 2007 protest letter about OU. Along the way, they lost Cyprus as an ally because the Commission had promised Nicosia that as a small and isolated market it could have a derogation. Nonetheless, on 29 January 2008 the band of eight tabled their proposal for what they called 'effective and efficient unbundling':

- The 'effective' part related to organizational independence of the transmission system operator (TSO) subsidiary from its parent group. This was to be achieved by ensuring the TSO had all the necessary assets to carry out its network responsibilities and would not be a shell company sub-contracting work back to its parent; that the TSO should have a different auditor to its parent company; that the TSO have a separate board with several independent non-executive directors; that a TSO chief executive could not go on to work for the parent company for a certain period of time; and that the TSO have a compliance officer to ensure these and other rules were obeyed.

- The 'efficient' part related to grid investments and market integration. TSOs would develop a 10-year investment plan, but if for some reason they did not carry it out, national regulators could step in to force them or third parties to do so. This was designed to make it hard to under-invest for the purpose of cordoning off home markets. The Commission might, at the request of TSOs, appoint a regional coordinator to promote cross-border interconnections.

The Commission's immediate reaction was that the Franco-German Third way would not lead to effective separation of networks from supply/production. But a more open attitude soon prevailed. It began to dawn on the Commission that, once the group of eight had signed up to a public position, it would probably stay intact as a blocking minority within the Council of Ministers. Thus it could stall any progress until spring 2009, at which time the EU's other co-legislator, the European Parliament, would be dissolved ahead of the mid-2009 elections. So at least a year might be lost, with no surer prospect of getting ownership unbundling through in 2010, and all the other proposals in the Third package would be jeopardized. So the Commission began to work together with Slovenia, which held the EU presidency for the first half of 2008, towards a compromise.

But the anti-OU camp suffered a setback, when, on the morning of 28 February 2008, Energy Ministers Council, Eon and the Commission announced that they had come to a preliminary deal in which Eon would sell off its German electricity grid and Brussels would drop its anti-trust investigation into alleged power market manipulation by the German utility. Peter Hinze, Germany's economics minister, learnt about this as he walked into the Justus Lipsius council building in Brussels on 28 February. He rounded on the Commission for its 'very questionable game', effectively accusing it of timing the announcement to humiliate him and weaken his bargaining position.[4] The Commission

---

4   Report by Reuters, 28 February 2008.

denied any such motive on the deal's timing, which it claimed
was dictated by the pace of negotiations with Eon. As it happened, however, announcement of the very similar deal on gas
between Germany's RWE and the Commission also came just
before the June 2008 Energy Council ministerial meeting.

The U-turns by Eon and RWE did not alter the stance
taken in council meetings by German ministers, who were also
angry at being made to look foolish by their companies. As
the practical reasons for defending their companies from OU
slipped away, German ministers elevated opposition to OU
even more to a matter of principle. But the sight of Brussels
forcing Germany's two biggest utilities into plea bargains over
unbundling had an impact of some of Germany's allies. 'The
Eon announcement was very important', said a senior Commission official some months after. 'It changed the dynamic because
whatever the German government said, there was no longer a
unified German position.'

The essence of the June 2008 compromise was that it contained an option for an 'independent transmission operator'
(ITO), allowing the network management to stay within the
integrated parent group, as the French and Germans wanted.
Nor was it treated as an inferior option, but given the same status
as OU. In a further concession to the Franco-German camp,
ITO status could apply to electricity as well as gas.

What did the opponents of OU concede in return? Essentially to wrap themselves in red tape and conditions and
to authorize national regulators to impose substantial fines
– which never figured in the Commission's original proposal
– on parent companies and network subsidiaries found guilty
of discrimination.

Among the main conditions were that an ITO must have
sufficient financial, physical and human resources so that it does
not need to contract everything back to its parent company. Top
management could move from the parent to the network, but
only after 'cooling off' periods of three to four years to prevent
people taking commercially sensitive information with them.
Parent companies could crown their ITOs with a supervisory
board, but nearly half these board members had to be subject
to the 'cooling off' restrictions, and national regulators could

step in to prevent any board member being 'unfairly dismissed'. Regulators must authorize ITO investment plans, and could force changes in them.

This compromise is likely to be the basis of the Third package that is expected to pass into law in spring 2009. In autumn 2008, the outcome was still complicated by the schizophrenia of the European Parliament, which had in its first votes sided with the Council of Ministers' gas compromise but on electricity went for ownership unbundling pure and simple. But the prevailing mood among MEPs as well as governments was a desire to get the Third package of energy market reforms on to the EU statute book, and to 'change the subject' to the other energy issues of security of supply and of climate change.

Speaking after the basic compromise reached at the June 2008 energy council, energy commissioner Andris Piebalgs said he had mixed feelings. 'Sadness, because the Commission proposal for ownership unbundling was a very good one. But political life in the EU needs member states to endorse proposals, and now that the way has been paved for final adoption of the package is a reason for satisfaction'.

For his successors as energy commissioner, a key task will be to ensure proper implementation of the Third package so that there is no need for a Fourth one. Yet even the Third package – for all its lengthy preparation – may have only had a narrow window of opportunity through which to slip on to the statute book. The deal on unbundling might not have been possible until the anti-trust investigations began to have an effect on German companies in spring and summer 2008. But by that time, France signalled clearly it would use its presidency of the EU in the second half of 2008 to focus on what it – and most other governments – regarded as more pressing business: energy security and climate change. And it is to these issues we now turn.

CHAPTER 8

ENERGY SECURITY: THE WEAKEST LINK

Europe managed to carry on sleep walking [into excessive dependence on imported gas] for many years before any alarm bells rang.

Paolo Scaroni, CEO of Eni speaking to the World Energy Congress, November 2007.

It may be valuable to consider proposing the creation of a more central role for the European Commission in the external energy relations of the EU, beyond its existing legal competences, by providing in particular for stronger coordination of member states.

The International Energy Agency, 2008 Report on EU Energy Policy.

When Paolo Scaroni talked of 'alarm bells', he was referring to Russia's transit disputes with Ukraine and Belarus of 2006–7, leading to brief interruptions in the westward flow of Russian gas and oil through those countries. He could not have known that in just over a year, Russia's military conflict with Georgia and two-week cut-off of gas through Ukraine would give its European Union energy customers a much more serious fright. Although the Russian–Georgian conflict was not motivated by energy, it had a big impact on perceptions about Europe's energy security. For it was seen to put the only energy corridor providing a non-Russian outlet for Caspian oil and gas right under the shadow of the Russian military. The implications of this will be dealt with in the next chapter. First, however, some general context is needed to understand what the EU can – and cannot – do to provide energy security for its members.

Energy security is a complicated issue, and becoming more so because of climate change. One day we may reach a stage where we still have enough fossil fuels, particularly coal, but are too scared to continue using them for fear of triggering climate catastrophe. Among low carbon energies, nuclear power raises issues of security in terms of safety: to some extent safe operation of reactors (which is much improved around the world), but relating more to the health hazards of radioactive waste. And the question mark over renewable energy is its security in terms of continuity, because almost all renewables are intermittent.

But, unless or until low carbon energy frees Europe from fossil fuels, it will need to import increasing amounts of them. For most Europeans, energy security boils down to a story of gas and Russia. First, gas, because its large-scale import of gas through fixed pipes creates a particular dependence on regional suppliers; other forms of energy are either largely domestic (renewables) or easier to import (coal and nuclear fuel) or tend to be worried about at more global level (oil). Second, Russia, because it is likely to be Europe's mainstay supplier for the future. It has the world's largest proven gas reserves, much of which happen to be within the reach of reasonable pipeline economics (maximum 3,000 Km) from the EU.

In this context, one has to say the EU has not, so far, been a very successful provider of energy security to its member countries. Are the latter better off than they would have been individually? Theoretically, yes, because they can help each other in emergencies even though in practice there is still insufficient storage of gas or interconnected pipes between them to make this a sure reality. Are the trade-offs between energy security and other energy policy goals such as emission reductions less constraining at 27 than as a single state? Theoretically yes, because if EU states in Central Europe can have their energy security anxiety about Russia assuaged by the Union, then they may be prepared to make more of an effort to combat climate change.

But the EU has not lived up to the fairly high potential it was given in Chapter 2 in providing energy security. The fundamental reason is that member states do not want questions of what energy they use, and where they get it from, decided at the EU level. This is why they have always insisted – even, as we shall see, in their latest constitutional effort, the draft Lisbon treaty – that issues of energy mix and supply remain a national prerogative. Outside suppliers know this, and are only too happy to exploit it. As a very senior Commission official, acknowledges, in deep frustration, 'as long as each member state is responsible for its own energy mix, the Russians are quite correct to address each country bilaterally'. Therefore, as two other Commission officials have written, 'Russia's perceived strategy [of divide-and-rule] is in fact the mirror of the EU's weakness, not the cause of it'.[1]

In these circumstances, all the Commission can do is to warn in general terms about rising import dependence, as it has done in successive green papers, and to urge governments to speak with a single voice to outside energy suppliers. The Commission cannot buy gas; companies do that, and if these companies were to band together and negotiate on quantities and prices in unison, the Commission would normally be duty bound to prosecute them as a cartel. Commission diplomats can roll out the political red carpet for petro-power potentates visiting Brussels and sign memorandums of understanding with them. Indeed, a recent round of Commission energy diplomacy designed to solicit potential gas supply for Europe's south east 'energy corridor' produced offers of 2 billion cubic meters (bcm) a year from Egypt, 5 bcm from Iraq and 10 bcm from Turkmenistan. But the Commission cannot build the pipelines to actually get any of this gas to the EU. That is for companies to do.

In three particular areas of energy security, the Commission has tried to carve out an EU role. The results have been modest. The EU has tended to play second fiddle to the International Energy Agency on oil stocks, has yet to arrange sufficient gas sharing or interconnections for emergencies, and has failed to exploit, in the energy field, its particular aptitude for international legal frameworks that has borne fruit in the climate change arena.

Policing stock levels

EU dependence on fossil fuel imports is forecast to continue rising fast: in gas from today's level of 57 percent of consumption to more than 80 percent by 2030 and in oil from 82 percent to

---

1   Christian Cleutinx and Jeffrey Piper, in *Pipelines, politics and power*, Katinka Barysch, ed., Centre for European Reform, 2008, p. 26.

93 percent over the same period.[2] The EU first passed legislation in 1968 requiring its member states to hold minimum stocks of oil. This legislation has been largely eclipsed by the emergency stock holding and sharing rules of the IEA, which was set up in 1974 and to which 19 EU states belong. But the EU's rules are still the primary legislation in this field for its eight non-IEA members. The Commission would like to update these rules, but only to bring them more in line with those of the IEA.

There is no comparable EU (or IEA) requirement for gas stocks, because until recently there has been less panic about gas cut-offs and because gas is much harder to store than oil. Indeed only a few EU states have the right kind of caverns to store it underground in any quantity. A 2004 directive on security of gas supply required member states to prepare national emergency plans in the event of a gas shortage, suggested they might like to store gas if they had the right geology, allowed bilateral gas-sharing schemes between member states, and set up a Gas Coordination Group to propose EU-wide measures in the event of 'a major supply disruption'.

This was defined, in the directive, as an interruption in which the EU 'would risk to lose more than 20 percent of its gas supply from third countries for a significant period of at least 8 weeks'.[3] Under that definition, no fewer than nine member states, mainly smaller East European ones, could lose their entire gas supply without the EU as a whole lifting a finger. Nor, in this absurd directive, was there any minimum level of gas stocks to give meaning to consultations.

The original 2002 proposal of the Commission was that each member state must ensure that it would have enough gas to maintain supply to 'non-interruptible' customers (mainly households with no possibility of switching to other fuels) if the single largest source of supply to that country were to be disrupted for up to 60 days. This was to be achieved through storing gas or drawing on a surge in production or having emergency sharing arrangements with neighbouring states. The proposal

---

2   Commission communication to the European Council, COM (2007) 1 Final, 3.

3   Directive 2004/67/EC.

also committed member states to gas sharing arrangements, over which the Commission would have had some discretionary power.

But the European Parliament complained this proposal gave public authorities, especially the Commission, far too much power over gas stocks which industry was quite capable of managing, and that in general the risks to gas security were far less than to oil flows. Together with the Council, the Parliament heavily rewrote the directive so that the only precision, in the part dealing with 'a major supply disruption', was not about gas stocks but about the circumstances in which the member states and Commission would meet to consult. In this way, the EU totally failed to cater for the particular energy security concerns of the group of East European states just about to enter the EU in 2004.

Holding dialogues with suppliers

Energy features on the agenda of many dialogues the EU has with countries or groups of countries, and is particularly prominent in meetings with groups involving Gulf, Central Asian, North African states, and of course Russia (discussed in the next chapter). But with only one group, Opec, is energy totally central to the dialogue.

Brussels' pleas for Opec to supply more oil in order to moderate price rises are generally as ineffectual as all other such pleas to Opec from oil consumers. Yet at a time of fluctuating oil prices and changing energy policy in Europe, the EU-Opec dialogue can serve a purpose. At their meeting in Brussels in June 2008, EU commissioner Andris Piebalgs made a point of saying that, even if all the EU plans to move to a low carbon economy materialise, 'the EU will be importing more oil in 2030 than we do today.' His words were directed at repeated Opec complaints over the previous year that EU plans to develop renewables, especially biofuels, would kill demand for the oil cartel's staple product.

For his part, Dr Chakib Khelil, the Algerian oil minister and then president of Opec, suggested that while EU plans for renewable energy and energy efficiency would lower demand for oil, they are actually less destructive of oil demand than the then current record oil price of over $140 a barrel. In the long run, the Algerian minister said: 'Opec member states will benefit from EU experiences – and technologies – in conservation, energy efficiency and carbon abatement, and we will encourage the EU to go in this direction.'[4] This is an example of where EU energy diplomacy can improve the political atmosphere with suppliers, though without any proven beneficial effect on volume or price.

Exporting policy

The counterpart of the EU's need to import physical energy has been an attempt to 'export' energy policy, meaning the political, commercial, legal terms on which the energy is imported. The first, and most ambitious, effort to do this was the Energy Charter Treaty. Dreamed up by a visionary Dutch presidency of the EU in 1991, as the Soviet Union was imploding, it was designed to create a legal framework for cross-border investment and trade in the energy sectors of the countries of the collapsing Soviet empire, though it was dressed up as a wider international agreement and has more than 50 signatories. Indeed it remains a landmark because it is the only specific international agreement covering energy; the World Trade Organization, and the Gatt before it, has no special rules on energy. The essential aim, for the EU, is to protect investment and to ensure that transit countries do not charge a ransom for letting energy pass through their territory.

But Russia never ratified the Energy Charter Treaty (ECT). So it is not binding on the one country it was most designed to apply to, thereby reducing the ECT to a case of Hamlet without the prince.

Subsequently, the EU tried another approach in another region. In 2005 it set up the Energy Community to include the western Balkan ex-members of the old Yugoslavia, Albania and,

---

4   Press conference, Brussels, 24 June 2008

until they joined the EU in 2007, Romania and Bulgaria. The original aim of this Community – which obliges its members to accept EU energy market principles and decisions – was to make post-war Balkan reconstruction more effective (through better organized cooperation) and to prepare states for eventual EU membership. But some in Brussels came to see it as a valuable potential bridgehead for exporting EU energy market policy further east, to Turkey and beyond.

Turkey's participation is important to EU energy diversification and any construction of new non-Russian pipelines. But other countries with no prospect of EU membership are unlikely to adopt the EU energy rules. For the wider region, the EU has tried to come up with a less onerous regime. In 1995, it created the Interstate Oil and Gas Transport to Europe (Inogate) programme of energy policy cooperation with Black Sea and Caspian states. An attempt to revive it was made in 2004 by rebranding it the Baku Initiative. But without the carrot of EU membership to dangle before participants, the only other incentive is money. And, as an official in Brussels remarked, 'the idea you can buy influence with small amounts of money is illusory.'

Diversity of supply

If EU countries are fixated on the problem of depending too much on one supplier, it is worth bearing in mind that that supplier is not always Russia. Spain is one of the few EU states to import no gas from Russia. But it imports a third of its gas from Algeria – 13.20 bcm out of total 36.13 bcm imports in 2007.[5] In fact, Spain is unique among EU states in putting into law in 2000 a self-imposed limit on the share of its total gas consumption coming from any one country. This limit was set at 60 percent, which in 2000 was the share of its gas coming from Algeria. (During the Cold War, Germany's Ruhrgas used to informally try to keep imports of Soviet gas to 30 percent of total West German gas, in order to quiet US fears about West German energy dependence on the Soviet Union).

---

5   BP Statistical Review of World Energy, June 2008.

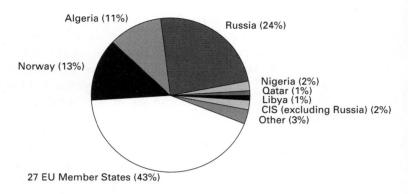

Figure 2: Diversity of Gas Supply

Source: International Energy Agency report on EU energy policy 2008,
         p. 62

Algeria sells far more gas to Italy (24.53 bcm in 2007), but
this was not much more than a third of Italy's much bigger
import volume (74.88 bcm in 2007). Algeria has plans to increase
overall exports from around 62bcm in 2006 to 85bcm a year
by 2010–12, and most of this increase is to be piped to Italy
(6.5bcm via the Transmed pipeline and 8bcm via the Galsi
pipeline to Sardinia). But Italy is due to match this increase
with extra purchases from Russia, keeping its reliance on Algeria
relatively unchanged. Several other EU countries – France,
Belgium, Greece and Turkey – also take Algerian gas. But they
take it in the form of LNG shipments (a form of transporting
gas pioneered by Algeria's Sonatrach). So they are less locked in
to Algeria than are Spain and Italy with their fixed pipelines.

The EU's third main outside supplier, Norway, exported 89.7
bcm of gas in 2007. This volume is expected to rise substantially in the next few years, taking Norway's total production to
118bcm in 2011. But this is more the step-change from bringing
the huge Ormen Lange field into production rather than any
intent to go on raising output. With only less than 5m people,
Norway needs no extra revenue.

Even if it wanted to, Norway is no position to exercise any
undue market power on the EU. For the country has signed up
to the European Economic Area, obliging it to make its practices

comply with Brussels' antitrust rules. One of these practices was Oslo's requirement that all gas producers in Norway had to coordinate their sales through a single organization, known as the GFU by its Norwegian initials. The Norwegian government's rationale for having a single selling organization was that this made its (cautious) depletion policy easier to operate. But Brussels insisted the GFU was an illegal cartel, and ordered it to be disbanded in 2001.

Spectrum of insecurity

No EU state feels entirely comfortable about energy security, because all except for Denmark are net importers of oil, and all but Denmark and Netherlands are now net importers of gas. But the degree of (in)security a country feels revolves around its dependence on a given type of energy and the number of its sources for that energy.

By these two criteria, France has, in terms of gas, reason to feel the most secure of EU states – and probably more so than the UK and the Netherlands which are relatively bigger users of gas because they have had their own (now declining) production. Not only does France use relatively little gas (half that in the UK and nearly as little as the Netherlands). But thanks to a mix of geography and policy, France, or Gaz de France, gets this gas from no fewer than fifteen different sources and by various routes, with each of its four main suppliers – Norway, Russia, Netherlands, Algeria – each supplying within 14–21 percent of total French imports.

At the other extreme of insecurity spectrum are the Central European countries, which for the most part were part of the Soviet Union or its empire and were therefore linked in to its gas system. So in addition to Finland, all three Baltic states are still 100 percent dependent on gas from Russia, as are Slovakia and Bulgaria. Austria, Czech Republic, Hungary, Poland and Greece get more than half their gas from Russia.

All other EU countries are somewhere in between France and Central Europe on this spectrum. The UK, Italy and Germany are relatively big users of gas but draw it from several sources.

By contrast, Poland is a country that has clamoured about the potential insecurity of its gas imports that come mostly from Russia. So one would guess that Poland is in the same position as other Central Europeans. In fact, it consumes one third of the gas that Spain, a country with the same population, uses; the main reason is that Poland generates 95 percent of its electricity with its own coal.

Some fellow Central Europeans believe that Poland and some of its neighbours have lapsed into a passive 'victim mentality' on energy security. One of these is Vaclav Bartuska, the Czech Republic's special ambassador for energy security. He recently wrote that 'most of the central and East European countries did very little to lessen their energy dependence on Russia. They have had plenty of time since the collapse of communism in 1989–91. But governments and people in the region mostly felt that joining NATO and the EU would bring sufficient security in all areas.'[6] Mr Bartuska contrasts this passivity unfavourably with his country's early and proactive policy of building, well before the Czechs joined NATO or the EU, an oil pipeline to Germany and contracting to buy some gas from Norway.

Yet Warsaw's worries, even if sometimes more imagined than real, about its energy dependence on Russia have had a more lasting effect on EU energy policy than Prague's pragmatism. Poland first made its proposal for an 'energy solidarity' commit-ment in 2006 to NATO, a forum where it knew it could count on support from the US, which had been seeking to push energy security issues higher up the alliance agenda. But the response from NATO as a whole was rather negative. Some members such as France felt NATO should stick to military matters and others such as Germany were wary of the reaction of Russia, which had made it clear it did not want to talk about energy security with NATO.

So Poland took its energy solidarity idea to the EU, and threw it into the Union's constitutional negotiations. It surfaced during the June 2007 summit negotiations to patch up a new EU treaty. Nominally, Lithuania led the push for a solidarity clause in the treaty. For Lithuania has a triple dependence on Russia: for

---

6   Vaclav Bartuska, in *Pipelines, politics and power*, p. 57.

gas, for oil (cut off for much of 2006–7), and for uranium fuel (for its huge Soviet-era nuclear power station at Ignalina). But what got everyone's attention was Poland's tactics in support of Lithuania.

Poland was the last country that the then German presidency of the EU had to square in order to get a deal on the Lisbon treaty. The Kaczynski twins, then running Poland as prime minister as well as president, used brutal rhetoric in support of Poland's quest to keep more Council votes than its population would normally merit. This included blaming the Nazis for the paucity of Poles relative to Germans today and therefore for Poland having too few EU votes relative to Germany.

In the circumstances, an energy solidarity clause was an obvious sop for Chancellor Angela Merkel, as EU president, to placate Poland. (Germany would, by size and location, be the most practical provider of energy solidarity to Poland, and luckily Berlin's relations with Warsaw rapidly improved after the defeat of Jaroslaw Kaczynski as prime minister).

So the draft 2007 Lisbon treaty has, under the heading 'Difficulties in the supply of certain products (energy)', a new sentence to the effect that: *'The Council, on a proposal from the Commission, may decide, in a spirit of solidarity between the member states, upon the measures appropriate to the economic situation, in particular if severe difficulties arise in the supply of certain products, notably in the area of energy.'*

The notion of solidarity also figures for the first time in the main reference to energy in the EU's putative new treaty, as does the creation of formal EU competence to *'ensure security of supply in the Union'* (see Chapter 1). But we have also seen that there is a crucial qualification that the treaty *'shall not affect a member state's right to determine the conditions for exploiting its energy resources, its choice between different energy sources and the general structure of its energy supply'*. This language could void the new wording about energy security of real meaning. And probably Poland would be as loath to surrender the national prerogative over energy mix as any member state.

Can formal treaty articles create solidarity? After all, the economic interdependence and constant political interchange of EU states today is such that they could hardly let a cry for help from one of their number go unanswered. But they have in the past. Such a clause might have been useful during the 1973–4 Arab oil boycott, which was particularly directed at the Netherlands, as well as the US, for being pro-Israeli. At that time, countries such as Belgium, joined to the Dutch through the Benelux organization as well as the European Community, gave way to Arab pressure not to pass on oil to their Dutch neighbours. Yet it is likely that policy on energy security will evolve – even more than other aspects of EU energy policy have – in response to necessity and events rather than to treaty wording. The question therefore is whether the Georgian conflict of summer 2008 and the gas cut-off of early 2009 mark the point at which the EU got serious about diversifying energy imports away from Russia.

CHAPTER 9

MANAGING RELATIONS WITH RUSSIA

The president stated he learnt that in Europe Russian gas is resold at three times the export price charged by Gazprom... 'Why is Gazprom supplying so cheaply, and, even more importantly, where is the price difference going, where is the money?' exclaimed Putin with indignation.

Report in Kommersant 21 November 2001 on Putin's meeting with Gazprom leadership.

Russia's energy policy follows a tight script, it has a sense of strategic purpose. There is, in principle, nothing that stops us, the Europeans, from matching their determination with our own discipline.

Javier Solana, High Representative for common foreign and security policy, February 2008.

Russia is the EU's most important, but also most difficult, energy partner. It provides nearly half (42 percent in 2005) of the Union's gas imports, a third (32 percent) of oil imports, a quarter (24 percent) of coal imports, and almost all of its enriched uranium fuel imports. Especially in gas, this dependence may grow for the obvious reason that Russia has the world's largest gas reserves and is still the main outlet for gas from central Asia. Moreover, the only outside electricity link that the EU's three Baltic states have is to the Russian grid.

Relations with Russia have not grown any easier as a result of the mass entry into the EU in 2004 of central and East European states, many of whom still carry a strong anti-Russian animus from their days as forced members of Soviet institutions and alliances like the Warsaw Pact and Comecon.

They evidently thought they would find greater energy security in EU membership than they did. As smaller states, too, these countries place a higher value on the Union speaking with one voice in its energy foreign policy, provided it is a tough voice. For instance, in September 2007 the European Parliament approved a report, written by the Polish chairman of its foreign affairs committee, Jacek Saryusz-Wolski. It called for a stronger and more collective 'common European energy policy, covering security of supply, transit, and investment related to energy security', and for the naming of 'a High Official for Foreign Energy Policy' to coordinate such a policy.

This *ostpolitik* of the EU's new members has brought them into disagreement inside the EU with the larger and older member states which have tended over the years to cut their own energy deals, or pursue their own political lines, with Moscow. But it is probably a mistake to think, as many in America as well as in Europe do, that the differing approaches of EU member states towards Moscow stem largely from their differing energy reliance on Russia. According to this widely held theory, energy actually determines the various EU views of Russia, ranging from the East Europeans' desperation to rid themselves of near total dependence on Russia to some bigger west European states' reluctance to offend Moscow in any way that could jeopardize their sizeable and lucrative bilateral energy trade and investment links with Russia.

Important people believe this. Successive US administrations have voiced concern that its European allies risk, by becoming too dependent on Russia for energy, exposing themselves to political leverage from Moscow. This is why the US has sought to raise the energy security issue in NATO, and why the EU has to endure a certain amount of 'diplomatic back-seat driving' from Washington on the matter of energy security. (In recent years the State Department has put a series of senior officials in charge of 'Caspian regional energy diplomacy'. An early part of this mission related to US energy security – building the Baku–Tblisi–Ceyhan oil pipeline to take Caspian oil to the Mediterranean and the world market, thereby influencing the world oil price that affects the US, like any other oil importer. But the rest of this State Department mission relates to trying to introduce into Europe Caspian gas that would have no impact whatever on North America's regional gas market. The US aim is to encourage energy diversification for both Caspian energy producers and European importers, especially in Central Europe.)

In fact, Germany's desire for a smooth relationship with Russia is based as much on the memory of the terrible damage the two countries did to each other in the Second World War, as on a historic symbiosis of Russian raw materials feeding German industry. France's cultivation of Russia goes back to General de Gaulle's foreign policy and earlier, predating any French import of Russian gas. Andris Piebalgs, the EU energy commissioner and himself a Latvian, believes that 'energy is sometimes being used as an excuse to hide' political reasons for taking different approaches to Moscow.[1]

Indeed, such is the multiplicity of political factors in relations with Russia that one observer claims to be able to divide EU states into five categories according to their attitude towards Russia.[2] These are 'the Trojan horses' (Cyprus and Greece); the 'Strategic Partners' (France, Germany, Italy, Spain); the 'Friendly Pragmatists' (Austria, Belgium, Bulgaria, Finland, Hungary, Luxembourg, Malta, Portugal, Slovakia and Slovenia; the 'Frosty Pragmatists' (Czech Republic, Denmark, Estonia, Ireland, Latvia, the Netherlands, Romania, Sweden, and the UK); and the 'New Cold Warriors' (Lithuania and Poland). Energy is not the common dividing line between these groups.

All this led to Peter Mandelson to say, when EU trade commissioner, that 'no other country reveals our differences as does Russia.'[3] Differences over Russia are internally very disruptive within the EU. The way that member states diverge in their political reactions to Russia is not totally unlike the way their currencies used to diverge in reaction to the gyrating movements of the US dollar; Germany's D-mark was always firmer against the dollar relative to other EU currencies (just as, some would say, Germany's political response to Russia is always softer). The response to monetary gyrations was to create a common currency. Nothing so mechanistic is possible in creating a common policy towards Russia. But a unified approach on energy

---

1 Interview with Argus Media Group, September 2008.
2 Mark Leonard and Nicu Popescu, European Council on Foreign Relations, 2007
3 Peter Mandelson, speech in Bologna, 20 April 2007.

would be an important building block towards a common foreign policy. And as a step to that it is important to look beyond perceptions to some realities.

Russian realities

In certain ways, Russia has become a more difficult energy partner than when it was the Soviet Union. For Western Europe, the politics of dealing with the Soviet Union were relatively simple. The Cold war, although it had its thaws and chills, kept these relations in a relatively steady state. Moreover, west Europeans knew where they were with the Soviets, who were clearly the enemy. But they were not beyond the pale when it came to energy. The first supplies of Soviet gas came to Austria (then neutral) in 1968, and two years later West Germany's Rurhgas signed its first long term contract for Siberian gas. Such was the confidence that west Europeans had come to have in Moscow as a supplier by the early 1980s that they were willing to defy the attempt by their major ally, US president Ronald Reagan, to curtail their dependence on Soviet gas.

But the collapse of the Soviet Union ushered in a more confusing era for Russia's neighbours (as well for Russians themselves). Russia went first, under Boris Yeltsin, through a very pro-western phase. During this phase it entered the fold of the democracies, was admitted to the Group of Eight, and subscribed to high-minded organizations such as the Council of Europe with its stress on democracy and human rights. But just as west Europeans started to feel they had acquired a perfect right, through organizations like the Council of Europe, to comment on Russia's political imperfections such as its brutal subjugation of Chechnya, Russia itself began to bridle at such 'interference' as the president it elected in 2000, Vladimir Putin, pledged to restore law, order and his country's fallen prestige.

Vladimir Milov, president of the Institute of Energy Policy in Moscow and an outspoken critic of Mr Putin's, charted this policy's effect in the energy field when he spoke to the European Parliament in February 2007. After 2003, according to Mr Milov, government policy moved away from privatizations towards greater interference in the economy and energy sector and arbitrary use of the state's regulatory powers to advance the position of state-controlled 'national champion' companies. The most spectacular example of this was the deliberate destruction of the Yukos group of Mikhail Khodorkovsky, whom Mr Putin considered a political challenger; he ended up in a Siberian jail for tax evasion. Whether or not Yukos or Mr Khodorkovsky were guilty of tax evasion, they were never given a chance to redeem themselves. Whenever Yukos appeared to come near paying a demand for back-tax, the authorities would hit it with a larger demand and/or freeze its bank accounts. The resulting dismemberment of Yukos provided a convenient occasion to bolster the assets of state-controlled Gazprom and Rosneft. The latter became the largest oil company, giving back to the state (which had never relinquished ownership of the gas industry) direct control of around 30 percent of the country's oil production.

Somewhat subtler but no less firm methods were used to squeeze some western majors out of upstream positions they had acquired in days when Russia was more receptive to foreign capital. Mr Putin particularly disliked what in June 2007 he described as the 'colonial' nature of production sharing arrangements (PSAs) that a few of the western majors had negotiated in Russia during the Yeltsin decade of the 1990s. His 'colonial' reference related to the fact that traditionally, PSAs, a form of financial ring-fencing to protect a project from arbitrary political or fiscal interference, were used in developing countries that might be considered unstable. At all events, the authorities used Royal Dutch Shell's alleged environmental shortcomings as a means of getting it to cede to Gazprom its predominant position in one of the PSAs on the island of Sakhalin off the Russian Pacific coast. It is unclear whether the Kremlin had any hand in causing the major problems that BP has had with its Russian partners in their TNK-BP joint venture, but Gazprom has had official support in moves to prise certain Russian gas assets away from the UK oil major. At the same time, Russia has passed more restrictive legislation on future foreign participation in strategic oil and gas fields.

Yet, as Mr Milov commented, 'Russian companies wish to buy energy assets in Europe freely.' It was, he noted, 'a strange concept of reciprocity'. It is mainly Gazprom that would like to buy direct sales companies in the EU, in the somewhat misguided belief that it has been missing out on enormous profit margins by using middlemen in the EU market. After meeting distribution costs and taxes, these profits margins are probably not very great. Nonetheless, as the quotation from Mr Putin at the start of this chapter makes clear, the gas company has come under pressure from its state owner to increase its profit and presence in the EU market.

But the main question marks about Russia's reliability as an energy partner have not arisen out of doubt about its attitude to contract sanctity – oil producers around the world have been renegotiating contracts as the oil price rose – but about Russia's ability to ensure continuous delivery. For, in contrast to the Soviet era, Russia no longer controls the transit routes for its energy exports.

Russia has always been the most land-locked of the world's big energy exporters. Though the world's largest country, its fossil fuels deposits have generally been far from the sea. This did not matter when Russia was still the Soviet Union, because all the transit routes – Ukraine and Belarus – as well as other oil/gas provinces – Azerbaijan, Kazakhstan and Turkmenistan – were also Soviet republics. Once the Soviet Union split up, transit was bound to become more complicated as Gazprom began to phase out subsidized prices to Soviet republics that were drifting out of Moscow's political orbit. Two incidents particularly spooked the EU. The first was Gazprom's cut-off of gas to Ukraine from 1–4 January 2006 as the result of a price dispute with Kiev. Then, exactly a year later, came an oil tariff row between Russia and Belarus that occurred in the context of a similar gas pricing dispute between the two countries. Any lingering doubts about the shakiness of post-Soviet transit arrangements for Russian gas to the EU were dispelled in January 2009 when Gazprpom stopped all gas to and through Ukraine for two weeks.

These were just the sort of post-Soviet energy transport difficulties that the Energy Charter Treaty, referred to in the previous chapter, was supposed to deal with. Russia signed, but unfortunately never ratified, this treaty, which therefore is not binding on Moscow.

For the Kremlin and the Duma had come to view the ECT as an unfair treaty foisted on Russia in its pre-Putin decade of weakness. Their reluctance to ratify was only reinforced by the attempt in 2000 to start negotiations on a supplementary Transit Protocol, creating an international legal regime for energy transit across multiple borders with enforceable dispute settlement. Gazprom claimed that the protocol was a deadly threat to its interests, even though the protocol stops short of the EU practice of requiring mandatory third party access to pipelines. In reality, as Jonathan Stern has put it, Gazprom's dilemma with the Transit Protocol has been that 'while it would effectively provide international sanctions against any transit violations by Ukraine and Belarus, it would also open the door to uncontrolled transit [across Russia] of Central Asian gas to Europe'.[4]

Yet the obstructionism was not all on the Russian side. The European Commission played its part in sabotaging Russia's adherence to the protocol. Russian negotiators wanted any quarrel involving transit across an individual EU state, as any other party to the protocol, to be taken to mutually agreed international arbitration. But the Commission insisted that any dispute involving the territory of an EU member state must go to the European Court of Justice in Luxembourg. Naturally the Russians did not regard the EU court as a totally impartial referee.

If energy were not a vital commodity, one could imagine relying on the passage of time for EU and Russian negotiators to see sense, and for both the transit countries and the new members of the EU to get over their 'post-Soviet traumatic stress syndrome'. On this analysis, EU-Russian energy relations would eventually settle down as Moscow 'market-izes' its relations with Kiev and Minsk, and as Poland and other new EU states become less paranoid about Russia. But since both sides have viewed energy as a vital commodity, they were weighing, at least before the Russian – Georgian conflict, the following steps to improve the situation or change its parameters.

---

4 Jonathan Stern, *The Future of Russian Gas and Gazprom*, OIES, Oxford, 2005, p.138.

Improving the status quo

Some powerful people in the energy business argue this is the only option. One such is Paolo Scaroni, chief executive of Italy's Eni group. As quoted at the start of the previous chapter, he complained to the World Energy Congress in autumn 2007 that Europe had 'sleepwalked' into being 'very reliant on a small number of gas suppliers', partly because Brussels 'concentrated all its efforts into fine-tuning the internal gas market, without grappling with the growing external threats'. Yet, in the next breath, the Eni chief went on to argue that 'under these circumstances [of dependence], it makes sense for the EU to build and safeguard good and cooperative relationships with its main suppliers, and in particular with Russia, with which it has geographical, historical and cultural links deepened by decades of mutually profitable trading.'

But safeguarding, let alone improving, this relationship is increasingly difficult. There was a steady deterioration in the EU-Russia political relationship during Mr Putin's second presidential term. Most EU governments, including Chancellor Merkel's government in Germany, have criticised Mr Putin's oppression of his opponents and his orchestration of the selection and election of a successor. Any hopes of EU-Russian relations improving with the advent of President Dmitri Medvedev were dashed by the Russian-Georgian conflict. This growing divide at the political level contrasts sharply with the cosiness between some of the big energy companies on both sides.

A dramatic sign of this dichotomy came in the autumn 2006 round of contract renegotiation, in which Germany's Eon, Gaz de France, Italy's Eni, Austria's OMV all extended their long term purchasing contracts with Gazprom for no less than 30 years, until 2035–6. At the same time, Gazprom has not been shy about its desire to advance into the European downstream with direct sales to European consumers. In return it says it is willing to give European companies a stake in developing the Russian upstream.

Both sides say they would like reciprocity, but mean different things by it. For the Russians, and for the bigger EU energy companies, negotiated asset swapping – upstream positions for EU firms in Russia and downstream access for Gazprom in the

EU – constitutes a perfectly satisfactory form of reciprocity. The asset-swapping formula was set out by Andrei Denisov, first deputy foreign minister, at a conference in November 2007. 'The capital assets [in Russian oil and gas] should be controlled by national corporations which should be getting their shares in supply and transport companies abroad in exchange for relevant minority shares in these assets.' For their part, the EU authorities want a legal framework conferring the same kind of automatic access that Russian companies, like all companies, have to the EU's single market. In fact, EU authorities – in Brussels and in national capitals – do not at present have the same ability to control Russian investment in their downstream that Moscow has in controlling European investment in its upstream.

So the European Commission has – in the context of its internal energy market restructuring (see earlier chapters) – come up with a proposal designed to give it some negotiating leverage on Moscow. The Commission has proposed a ban on any non-EU company taking majority control of any EU energy network, unless the home government of the buyer has an agreement with Brussels allowing such a purchase.

The aim of this clause, dubbed 'the Gazprom clause', is to prevent non-EU companies circumventing any new EU rules on the unbundling of energy networks by, for instance, a Russian financial fund with hidden links to Gazprom buying a gas pipeline in the EU. This is a safeguard to ensure equality of treatment so that Gazprom could not just snap up any networks that EU gas supply companies might be forced to sell. It has been denounced in Russia as an attempt 'to dictate to Russia the way in which it should regulate the operation of its energy companies in its domestic market.'[5] This is not so. The Commission was not trying to force Gazprom to get out of the gas supply business in Russia, any more than it was forcing Gazprom to buy EU networks. All Brussels was saying was that the EU should have the means to check Gazprom followed whatever new network unbundling regime that EU gas supply companies will have to

---

5   Sergei Yastrzhembsky, (former Kremlin adviser on EU relations) Pipelines, *Politics and Power*, Centre for European Reform, 2008, p. 37.

obey. However, it is true that the legislation – which does not mention Gazprom by name and ostensibly applies to all non-EU companies – was drafted with Gazprom very much in mind. Indeed Commission president Jose Manuel Barroso said publicly, when the proposals were announced in January 2007, that the aim of 'the Gazprom clause' was to give EU negotiators some leverage in order to obtain a more equal form of reciprocity in a new round of EU-Russia talks.

In July 2008, the EU and Russia started negotiations aimed at replacing their 1997 Partnership and Cooperation Agreement (which ran its intended ten years, but continues thereafter in effect unless discontinued by either party or replaced by a new accord). The idea was to reach a broader agreement, with substantial provisions on energy. Because of the Georgia conflict the EU briefly suspended the negotiations.

These negotiations will be lengthy and tough. EU states have a long laundry list of political issues on which they, probably unrealistically, want concessions from a Russian government that is relatively happy with the status quo in EU-Russian relations. But, to its own disadvantage, Moscow has contributed to the EU desire for change. The proposed restrictions on foreign investment in the Russian oil and gas upstream, and Gazprom's push to expand in the EU downstream, have both had the effect of making Russia's version of reciprocity look lop-sided to Europeans.

Diversifying energy routes

Part of the security question mark hanging over Russia's gas exports to Europe stems from its inability to arrive at smooth post-Soviet commercial arrangements along its traditional transit routes through Ukraine, Belarus and the Baltic states. So, if the routes are the problem, why not change the routes? This is what Russia is doing – proposing new undersea gas pipelines and overland oil pipelines to deliver energy more directly to most EU customers.

Any strategy that multiplies alternative pipeline routes increases reliability of delivery and so energy security for the

European customer. This is why many of the big EU energy companies, and, to an extent, the European Commission, have become willing accomplices in this Russian plan of new pipelines to sidestep old problems. But this is not altruism on Russia's part. Multiple alternatives also increase Russia's bargaining power over its traditional transit partner countries.

Gazprom has long thought strategically in developing alternative pipelines. In bringing Siberian gas to Europe, it has supplemented the main trunk route through Ukraine–Slovakia–Czech Republic with another pipeline through Belarus to Poland. Not content with being able to play Ukraine and Belarus off against each other, Gazprom is now proposing alternatives to both.

The biggest and most controversial project is the North European Gas Pipeline, otherwise known as Nord Stream. This would take gas, up to 55bcm a year, from the St Petersburg area through the Baltic to northern Germany. This project is made up of Gazprom (51 percent), Eon and BASF (20 percent each) and Gasunie of the Netherlands (9 percent); through the latter's participation Gazprom has also acquired a 9 percent stake in the BBL interconnector to the UK.

Nord Stream has stirred particular controversy in Poland. For Warsaw, the only thing worse than having Russian gas crossing Polish territory (and thereby maintaining dependence on a politically unwelcome source) is not having Russian gas crossing Polish territory (and Gazprom having another means to get gas to its richest markets in Germany and points west). In 2006, before he became Poland's foreign minister, Radek Sikorski, likened the basic Gazprom–German deal underlying Nord Stream to the Ribbentrop–Molotov pact of 1939 (which allowed for the Nazis to invade Poland without fear of drawing the Soviets in against them). But there are also environmental worries raised by states around the Baltic, and by the European Parliament, about running a big double pipeline through an area of sea in which German chemical weapons were dumped towards the end of the Second World War.

Since diversity of energy route, as well as source, is a guiding principle of energy security, the European Commission has endorsed Nord Stream as 'project of European interest'. Asked whether he should be supporting a project that bypassed an EU transit state, Poland, EU energy commissioner Andris Piebalgs justified his stance by pointing out that it also bypassed Belarus.[6]

As a sort of mirror image to Nord Stream, Gazprom and Italy's Eni have announced the South Stream project to take gas across the Black Sea to Bulgaria, whence it would go either north through the Balkans to Hungary and Austria or across the Adriatic to Italy or both. South Stream would duplicate existing and planned pipelines that pass through Turkey – the existing Blue Stream from Russia across the Black Sea to Turkey; the new Interconnector Greece–Italy (IGI) that pipes gas from Turkey; and the Nabucco project to take gas from the Caspian region to Central Europe. So apart from reflecting Gazprom's general expansionism, Italy's rising anxiety about its geographic position as last-in-line for Russian gas, and doubts about Nabucco, another purpose of South Stream may be to bypass Turkey for fear it may become too greedy a middleman (see below).

Russia is planning a third transit avoidance pipeline, this time for oil, to bypass Belarus. It will take 1m barrels a day of oil and oil products from western Russia, near the Belarus border, 1,200 Km north to the Russian Baltic port of Primorsk. This project known as Baltic Pipeline System-2 was started shortly after Russia's energy tax dispute with Belarus in January 2007. But it will also have implications for the Baltic states which, while hankering for less dependence on Russian energy in the long run, have come to rely on the fees they get for allowing the Russians to transport oil to ports like Ventspils in Latvia and Tallinn in Estonia. At the same time, Russia has cut off pipeline supplies to Lithuania's Mazeikiu refinery (after it was sold to a Polish, rather than Russian, buyer) on the ground that the pipeline is too corroded; so Lithuania now imports the same Russian oil by sea, and at a higher price.

It is up to Russia to decide how it wants to export its energy. But its strategy of export route diversification has consequences for foreign partners and customers. Transit countries will find it harder to resist Russian pressure on them to charge lower transport fees. The impact on final customers may be double-

6   Interview with Argus Media, September 2008.

edged. On the one hand, most final customers in the EU (though not Poland) can have more certainty about Russia's gas getting through to them. On the other hand, Russia could draw more of other countries' gas into its widening net of export pipelines. Russia has made it very clear its determination to remain the main geographical export outlet for central Asian gas. One motive for Gazprom building Blue Stream across the Black Sea, from southern Russia to Turkey, may have been to lessen Turkey's desire to buy non-Russian gas directly from the Caspian and central Asia. Yet South Stream might one day provide a Russian-controlled conduit for central Asian gas. All this could increase the dominance of Russia's gas market position, and the temptation to abuse that dominance by perhaps raising prices. It is therefore eminently sensible for the EU to make efforts not to avoid, but to supplement, Russia as an energy supplier.

Diversifying energy supply

Europe is far better placed to do this than most other gas-importing regions of the world. It has ²/₃ of world's gas reserves within 3,000 Km of it, or ¾ of such reserves within 4,000 Km. This latter distance is a bit beyond the current outer limit of pipeline economics, but there is always ship-borne LNG for most EU states on or near a sea. However for others, especially the more land-locked Central European states, hope for diversification could come from developing a gas corridor to the Caspian and the Middle East.

The main scheme here is the Nabucco pipeline. The gas company executives who met in Vienna to create the project gave it this exotic name, simply because that was the Verdi opera they had all been to see the night before. (In fanciful hindsight, an opera about Israelites yearning to escape their Babylonian captivity could be seen as a metaphor for Europeans yearning to escape Russia's energy grip on them). The idea of this pipeline is to pick up gas in and around the Caspian, central Asia, Iraq, the Middle East as far east as Iran and as far south as Egypt, and to take it across Turkey up through the Balkans to Hungary and Austria. Running the project is a consortium

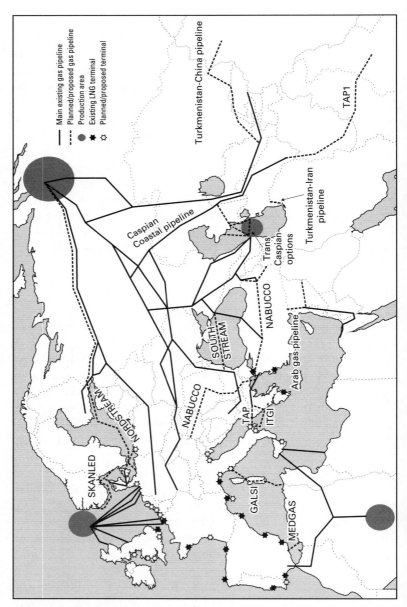

Map 1: Main Gas Pipeline Projects to Europe up to 2015

Source: International Energy Agency Gas Market Review 2008, p.58.

headed by Austria's OMV oil and gas company, with the four national gas companies from the other four countries along the pipeline route – Hungary, Romania, Bulgaria and Turkey – plus Germany's RWE. To underline its importance as the core of the EU diversification strategy, the European Commission appointed a special coordinator, the former Dutch foreign minister, Jozias van Aartsen, to help with the pipeline diplomacy involved.

However, two big question marks hang over Nabucco. One is where will the gas come from. Azerbaijan's Shah Deniz gas field in the southern Caspian Sea, which is already pumping gas to Turkey in a pipeline that parallels the Baku–Ceyhan oil pipeline, may be sufficient to get it started. But Nabucco will definitely need gas from other neighbouring sources. The most obvious place to look is Turkmenistan which has very big onshore gas reserves. So far it only exports to Russia (though shortly also to China), and the Russian government and Gazprom are keen to keep it that way. Russia has substantially benefited in the past from buying Turkmen gas cheap and selling it on dear to the Ukraine and Europe. In his May 2007 meeting with his fellow presidents of Kazakhstan and Turkmenistan, Vladimir Putin effectively persuaded these two countries to continue to funnel their gas through Russia. The Russian leader initialled an agreement whereby Turkmenistan would sell Russia an extra 30bcm a year. To do this, Gazprom had to agree to raise its price for Turkmen gas from $100 thousand cubic metres (a price that was earlier agreed to hold till 2009) to $130 during the first half of 2008 and $150 in the second half of 2008.

Under its new leader, Turkmenistan is potentially more open to the west. Yet in vying for its gas, EU leaders can hardly compete with single-minded Russian gas diplomacy for Turkmen gas. European commissioners not only cannot sign gas contracts; they are also under internal political pressure to protest about Turkmen human rights abuses. In that cause, the European Parliament has frozen negotiations for a formal EU agreement with Turkmenistan.

Turkmenistan says that it plans to quadruple gas production to 250bcm a year by 2030. But for the time being, it looks unlikely to have much to spare for Nabucco, after increased shipments to Russia and new ones to China. A big gas find in

Turkmenistan's offshore Caspian zone would greatly increase the economic rationale of piping the gas across the rest of the Caspian to Azerbaijan and thence to Turkey and on, via Nabucco. Even so the politics of doing this have been worsened by the events in Georgia. Speaking two months after the conflict, Mr Piebalgs was in no doubt about this. 'Russia has a huge interest in energy resources in this region. It is ready to buy all the gas from Turkmenistan, not to mention Kazakhstan, and to pay a good price.' The EU commissioner did not rule out central Asian states sending energy to Europe 'because diversification is important for these countries'. But he concluded that 'if I were in their place, I would be very cautious to take decisions that could complicate their relations with Russia.'[7]

The other doubt hanging over Nabucco is whether it can get reasonable transit terms from Turkey. For there are some worrying signs that Turkey could, in terms of rapaciousness for transit fees, become the Ukraine of the south. Some Turks see their country as constituting 'the fourth artery of Europe's energy supply after Russia, Algeria, and Norway' because of its ability to carry gas from many sources in central Asia, the Middle East and North Africa.[8] The phrase fourth artery is mirrored exactly in the European Commission's tendency to talk of Turkey as the 4th Gas Corridor. The bracketing of Turkey, a transit country, with three actual originators of gas is disturbing, and fuels fears that Turkey wants to be an arbitrageur – like Ukraine with Russian gas, or Russia with Turkmen gas – buying gas from the east cheap and selling it on to the west dear.

For precisely this reason, the European Commission has been keen to entice Turkey to join its Energy Community. Essentially, this community is a vehicle for extending the EU's existing legislation, or 'acquis' in the Brussels jargon, to neighbouring countries. The Energy Community rationalizes the transit issue that so bedevilled the Energy Charter Treaty out of existence. Once inside the Energy Community a country is part of a

---

7   Ibid.
8   Senior Turkish diplomat, quoted by John Roberts, in 'The Black Sea and European Energy Security', in *Southeast and European Black Sea Studies*, 2006, June, Vol. 6: 2, p. 216.

single (energy) market in which the concept of transit across a third country does not exist. Nor would, say, Turkey, once in the Energy Community be allowed to have dual pricing of energy by, for instance, raising the transmission price for exports and subsidizing down the domestic price. Inside the Energy Community, Turkey would have to charge the same cost-related fee to let gas traverse its territory as any state inside the EU with gas pipelines crossing it.

It remains unclear whether Turkey will sign up to the energy part of an EU that it may never actually join. If Ankara did sign up, then Commission officials have suggested that Turkey should be offered energy security guarantees, in the same way that some Balkan countries were provided with electricity from the EU under an 'energy for democracy' programme. Whether such energy security guarantees would be credible for a country the size of Turkey is another matter. It is inherently odd for the EU to offer such guarantees when it, the EU, is more a 'consumer' than 'producer' of energy security. On the other hand, financial guarantees backed by the EU budget or European Investment Bank loans to countries reorienting their energy production and transport infrastructure towards Europe might be conceivable.

Outside the energy sector, the EU has been very successful in exporting its laws, rules, norms and standards to other countries, especially to those neighbours with a real prospect of joining the Union eventually. But, where energy is involved and where there is doubt over a country like Turkey ever joining the EU, these 'policy exports' will have to be accompanied with something tangible like cash. That is the view of Mr van Aartsen, in his role as EU coordinator for the Caspian–Middle East–Gas Route. 'Some infrastructure projects are of such great importance that we should realistically expect some form of public subsidy for their realization', he wrote in June 2008. 'I would put Nabucco, or an equivalent route, in that category.'[9] That hope, however, was voiced before the general financial crisis of autumn 2008.

Future perspectives

The EU has every interest in preventing deterioration in the mutually beneficial energy relationship that some of its member states have had with Russia for over 40 years. But part of the difficulty is that Europe and Russia are two energy markets moving in opposite directions. The European Commission is trying to create another seismic shift of liberalization in the EU market, and nothing would suit Brussels or indeed European consumers better than to have Gazprom broken up into five or six gas companies all competing to export to Europe. Instead, Europe faces one huge vertically integrated monopoly in gas, and consolidation of the Russian state's ownership and control in oil.

Only in electricity is there parallel movement in the same direction. The privatization and unbundling of Russia's UES electricity system, carried out by its CEO, Anatoly Chubais, the one significant holdover from the Yeltsin years, matches what has been done in much of the EU. German, French and Italian companies have been encouraged to invest in electricity generation in the UES system. When faced with EU complaints about reciprocity and the lack of investment opportunities in upstream Russian gas, the Russian authorities can correctly point to the welcome they have given to foreign investors in their power sector.

Nonetheless, Russia has not lived up to the St Petersburg principles of 'Global Energy Security'. Vladimir Putin made these the hallmark of his year, 2006, as president of the Group of Eight industrialized countries. The St. Petersburg declaration on energy security included such principles as:

- *'transparent, equitable, stable and effective legal and regulatory frameworks, including the obligation to uphold contracts, to generate sufficient sustainable investments upstream and downstream'.*
- *diversification of energy supply and demand, energy sources, geographical and sectoral market, transportation routes and means of transport'* [10]

---

9   'European Union: The Energy Issue', *Financial Times* Business publications, June 2008, p. 54.

10   Declaration of the Group of Eight, St Petersburg, June 2006.

But within months of sponsoring this charter of economic liber-
alism and rule of law, Mr Putin was busy forcing companies out
of their 'colonial' contracts and enshrining in law Gazprom's de
facto monopoly on gas exports. The move of Dimitri Medvedev
from chairman of Gazprom to president of the country in 2008
further reinforced the identity of Gazprom with the state.

What can the EU do? It can hope one day to negotiate a
better legal framework on energy trade and investment with
Russia, as well as for warmer political relations. But it takes
two to achieve such results, while there are unilateral steps
the EU can take to improve its own energy security. In the
short term, it can, inside the EU, improve gas storage, create
gas-sharing arrangements and fill in the missing links between
national energy grids. In the medium term, the EU can take
the St Petersburg declaration to heart and begin to diversify
energy sources and routes. Nabucco would not have to be fully
built, merely underway, for Europe to feel some of the easing
of commercial terms that could come from diversification. Most
important but over a longer time frame, it can carry out its
low-carbon revolution to reduce Europe's dependence on fossil
fuel imports, from Russia and everywhere.

Energy commissioner Piebalgs produced the following projections in autumn 2008 to show this could be possible. If the oil
price were a relatively low $60 a barrel, keeping gas prices low
and demand up, and in the absence of the Brussels renewable
energy and climate change package, EU gas imports would rise
50 percent from around 300bcm to 450bcm by 2020. In the
absence of new climate change policies but a $100 oil price, gas
imports would fall to 380bcm. But gas imports could actually
fall below today's levels, to 280bcm in 2020 – if, in addition
to a $100 oil price, Brussels' new energy and climate change
package all worked as planned. This is a very big assumption.
Is it realistic or heroic?

---

11 Interview with Argus Media, September 2008.

CHAPTER 10

RISING TO THE CLIMATE CHANGE CHALLENGE

The needs of the environment are coming together with the needs of the EU: one is a cause looking for a champion, the other a champion in search of a cause.

David Miliband, UK environment secretary, November 2006.

Climate change was an obvious challenge for the EU to rise to. One essential reason is the consensus its member states share about the problem. This consensus is surely not unconnected with the projected geographical impact of unchecked climate change within Europe. This is expected to be harshest in terms of high temperatures and drought across southern Europe, and the map below (based on the Eurobarometer opinion poll) shows that concern is indeed greatest across the EU's southern belt. This region is also likely to receive even more economic refugees from the southern shore of the Mediterranean and sub-Saharan Africa both of which will be still worse afflicted by climate change. In the EU's continental heartland, in central and eastern Europe, summer rains are projected to decrease.

Yet northern Europe, the one area of the continent where global warming might bring some partial benefit with less need for winter heating and higher crop yields, also happens to be the EU's politically greenest region. Mainstream political parties in northern Europe are generally environmentally aware, and sometimes environmentalists hold the balance of political power, as the Greens have done in Germany. The Greens are also an important force in the European Parliament, where they are better represented, due to the EU-wide system of proportional representation for the Strasbourg assembly, than in those national legislatures operating on a first-past-the-post system that benefits incumbent mainstream parties. The UK Green party, for instance, has one MEP at Strasbourg but no MPs at Westminster.

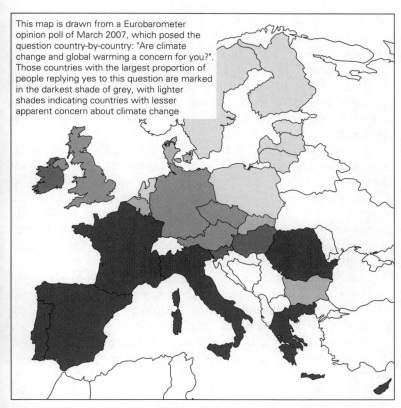

This map is drawn from a Eurobarometer opinion poll of March 2007, which posed the question country-by-country: "Are climate change and global warming a concern for you?". Those countries with the largest proportion of people replying yes to this question are marked in the darkest shade of grey, with lighter shades indicating countries with lesser apparent concern about climate change

Map 2: Mapping Climate Change Concern

Source: Eurobarometer March 2007

EU action on climate change also fits into a tradition of pre-emptive environmental measures that had grown up not in only in some important member states such as Germany and Nordic states, but also at the Union level. This tradition is encapsulated in the 'precautionary principle', the notion that sometimes you have to act early – even before you have conclusive proof of a problem – because the problem, left untended, could result in enormous and irreversible damage. EU states wrote this principle into their 1992 Maastricht treaty.

The treaty provision (Article 130R) said that: 'Community policy on the environment...shall be based on the precautionary principle and on the principles that preventive action should be taken, that environmental damage should as a priority be rectified at source and that the polluter should pay.'

Subsequently, the Commission published a paper in 2000 arguing that the precautionary principle should apply beyond the environment to all aspects of public health, and went on to put it into practice in the so-called REACH directive on chemicals.

The relevance of REACH (standing for Registration, Evaluation, Authorization and restriction of Chemicals) to the climate change debate is in the precautionary philosophy behind it. The 2006 directive was essentially a vast regulatory catch-up exercise on the large number of chemicals put on the EU market before 1981, when a proper EU-wide approval system came into existence. It requires all these older chemicals to be re-registered, re-evaluated and re-authorized, not because they are individually suspected of being harmful but because as the general class of pre-1981 chemicals they might be, and if so, they could cause hard-to-reverse biological or environmental harm. But this process is not cheap. Many companies, not least US firms, protested that the EU should wait for specific proof of harm before re-testing their products. Their protest echoed that of global warming sceptics demanding irrefutable proof, which can sometimes only follow irreversible change. In sum, what the REACH controversy showed was that in more than climate change has the EU been ready to follow the precautionary principle, to take a lesser risk (of, say, requiring unnecessary chemical tests) to avoid a bigger risk (of chemical damage to the environment).

The same precautionary approach was directly applied by the UK economist Nicholas Stern and his team in their 2006 report on the economics of climate change. It powerfully argued the case that 'the benefits of strong, early action considerably outweigh the costs', which it estimated at around 1 percent of global GDP by 2050. The report called this level of cost 'significant but manageable', and far less than the possibility of 20 percent reduction in global consumption per head if climate change were allowed to run uncontrolled. In fact, by the time the report came out, the precautionary argument was widely accepted in Europe. The report's main political purpose lay in trying to calm fears of those outside Europe – Americans, Chinese, Indians – about the economic costs of mitigating climate change.

Yet another reason why Europe has embraced the climate challenge may be to do with capability, a feeling that mitigating global warming by using less energy is not an impossible task for Europeans to achieve. Europeans generally do not have the vast distances, harsh climate and poor public transport that make energy saving in the US genuinely hard. But nor have they become anything like as oil-efficient as the Japanese, who now find further savings really hard. European countries mostly have a sufficient foundation of public transport networks and energy taxation policies to build credible climate change policies on. So they have no reason to despair of doing better, as sometimes Americans and Japanese, for opposite reasons, do. European industry has also become less energy intensive. This is true of all mature economies. But the change may be sharper within the EU, with the collapse of heavy industry in eastern Germany and some Central European states, and Britain's shift to a service economy.

But the reason for giving (in Chapter 2) climate change the highest potential rating of any energy-related EU policy is that the character of the EU creates an institutional match for the climate change challenge. Global warming is the ultimate cross-border problem. A task requiring an unprecedented degree of cross-border cooperation comes relatively naturally to an organization specializing in cross-border cooperation. If the EU had not existed, some kind of ad hoc 'green federation' might have had to be cobbled together to tackle the problem in Europe.

Yet it is because the EU is 50 years old, and has grown in numbers and ambition over that time, that it has been able to provide world leadership on this issue. This leadership role appears to have had public support. But there is nothing necessarily moral or lasting about this. In a Eurobarometer survey of March 2008, 64 percent of Europeans felt the environment was more important than competitiveness, compared to 18 percent who thought the opposite. This will not last as economic recession follows the financial crash of autumn 2008. People are bound to be more wary of the extra cost of renewable energy. Climate change will surely recede as a priority for companies, except perhaps insurers who always have to calculate the odds of extreme weather events happening.

Despite this, EU governments can take political courage from the fact that there is a certain amount of economic safety in numbers. Being part of a 27-country regional bloc, with a combined market of 500m people, means many EU states do most of their trade with each other. So they are not at a competitive disadvantage, as they all have to bear the cost of EU environmental initiatives. In international climate change negotiations, the 27 countries, with the Commission as their mouthpiece, carry weight that can be hard to resist. A frequent EU tactic is to argue, with some truth, that the common position of the 27 countries is the result of such delicate internal compromise that it cannot sustain any concession in subsequent negotiations with third parties.

On a smaller canvas, the Commission's ambitious January 2008 package is also the result of a shift in the attitude of the EU executive's president, Jose Manuel Barroso. One Brussels insider dates it from early 2006. 'He had heard Tony Blair talk very well about climate change at the Hampton Court summit in autumn 2005, and heard this echoed by French and German leaders. As a former prime minister, Barroso sees himself very much as one of the boys at European summits, and found climate change went down well there and that it was something he could discuss on equal terms with leaders outside Europe.'[1] Moreover, after voters in the French and Dutch referendums had killed the European constitution in 2005, Brussels was looking desperately for something to give the Union a lift. 'Barroso realized climate change was a good message to sell, that it fit well with his "citizen's agenda" and "Europe of results" slogans', said the same official. It also attracts younger generations, for whom the EU's post-war rationale of bringing peace to a ravaged continent tends to be meaningless.

Other commissioners too have backed stronger climate

---

1   Author interview, 2007

change policies – most essentially energy commissioner Andris Piebalgs, but also Jacques Barrot, who when he was transport commissioner consistently supported the inclusion of aviation in the emissions trading scheme (ETS). The exception has been Gunter Verheugen, the German commissioner responsible for 'enterprise and industry'. Because of his portfolio and his nationality, Mr Verheugen was bound to resist tougher vehicle emission standards; they bear hard on the heavier models made by German companies. So it proved. 'Getting the main climate change package through the Commission was a piece of cake compared to the car issue', said one official. In the end, it took a team effort, with Mr Barroso and the Commission's central secretariat knocking departmental heads together. 'Our challenge was to move beyond the environment, not to design something on the bright green extreme', said another official. The result was a complex package of proposals on the broadest environmental issue of our time, based on a lot of politics, economics and some lessons from recent mistakes.

Trial and error

Many EU governments combine high rhetoric about controlling climate change in the global interest with low tactics to further their national self-interest inside the EU. Almost every aspect of climate change produces behaviour termed as 'the prisoners' dilemma'. Each arrested prisoner knows that, if the optimum outcome (release for all) cannot be achieved by everyone pro-testing their collective innocence, then the next best thing is to be the first to denounce fellow prisoners. So it is with climate change policy inside as well as outside the EU: the next best thing to collective action is to be the first one to cheat. (Many Opec members take the same approach, publicly protesting solidarity with any collective production restraint to raise prices while quietly exceeding their oil quota to increase their benefit from higher prices).

Low tactics played a part in the initial mess the EU made of its emissions trading scheme (ETS), the main mechanism it set up in 2003 for implementing the greenhouse gas reduction targets laid down in the 1997 Kyoto Protocol. The EU had some cap-and-trade models to follow, chiefly the US sulphur dioxide and nitrogen oxide trading scheme, the carbon permit schemes started by Denmark and the UK, and some internal carbon trading schemes instituted inside BP and Shell. But the ETS was on a far bigger scale than these national or corporate efforts. And so were some of its mistakes, at least in the First phase of 2005–7.

The biggest mistake was to let national governments propose how many permits to allocate to their industries. As a result, the industries of some countries got more permits to emit more carbon than they were actually pumping into the atmosphere. This made nonsense of the cap-and-trade concept, which depends on the cap being a cut. There must be a shortage of permits in order for these permits to become worth paying for and trading. Once this over-allocation was revealed, in spring 2006, the price of CO_2 permits for the First phase of ETS crashed to virtually zero and never recovered.

The Commission was much tougher with governments about allocations for the Second ETS phase of 2008–12. Too tough, in fact, for five East European member states that went to the European Court of Justice with their complaint about stingy allocations from the Commission. Some carbon traders remain nervous about the Commission making another mistake, especially in the transition to the reforms to the Third phase of the ETS that Brussels proposes (see below). But the price of permits in the Second ETS phase has been fairly stable, though in decline with the recession.

Over-allocation has not been the only problem. Free allocation of permits has also caused distortions (auctioning has been allowed up to 10 percent of the total, but only about 1.5 percent of permits have actually been sold). Free allocation has given windfall profits to those companies – chiefly in the power sector – that have passed to customers the 'cost' (at the ETS market rate) of permits they received for free. There have been misallocations to new entrants. They have tended to be given permits to cover emissions for the entire production or generation of their new plants, thereby reducing their incentive to invest in low-carbon technology.

Nevertheless, the EU has made some progress towards its Kyoto targets. These targets are set at a collective eight percent reduction by 2012 (compared to 1990) for the fifteen countries that belonged to the EU at Kyoto's signature in 1997. It is this collective eight percent reduction target that counts for legal compliance with Kyoto, but within the EU–15 it is broken down into differentiated national targets, ranging from a 28 percent cut for Luxembourg to an increase of 27 percent for Portugal. Ten of the twelve subsequent EU members have individual reduction targets of six to eight percent (relative to various base years), while Malta and Cyprus have none.

Newer EU member states will generally have an easier time meeting their Kyoto targets, because many of them have accommodating base years, such as Poland whose base year for calculating its emission target was the relatively heavy carbon-emitting year of 1988. Meeting individual targets will generally be tougher for longer-standing EU members, and quite impossible for Spain. That country's economic boom lasted through the 1990s and beyond, 15 years in all. As a result, Spain's emissions are projected to be nearly 30 percent above 1990s levels by 2010, or double its Kyoto-derived target of a 15 percent increase.

For all the EU–15, emissions were 2 percent lower in 2005 than in 1990. This represents a substantial decoupling of emissions from growth in gross domestic product, which for the EU–15 increased by 35 percent in the 1990–2005 period. In theory, the consequence of missing a Kyoto target is painful. The protocol requires the over-shoot, plus a further 30 percent as a penalty, to be made up in the so-called 'Second commitment period' after 2012. In practice, the threat may be moot. A 'Second commitment period' has not been negotiated, and may never be, if the US and others insist on making a successor regime very different from Kyoto.

At all events, the EU would only suffer the Kyoto penalty if the 8 percent reduction target for all EU–15 were missed. And, notwithstanding Spain, this looks unlikely. According to Commission estimates in November 2007, the EU–15 should have reduced emissions by 4 percent by 2010, the mid-point in the Kyoto compliance period of 2008–2012. Add a further estimated saving of 0.9 percent from EU states' plans to plant more carbon-absorbing trees, and another 2.5 percent from plans by 10 of the EU–15 states to buy emission credits from outside Europe, and the EU–15 should get to a reduction of 7.4 percent. In addition, the tightening of national allocation plans in the Second phase of the ETS, 2008–2012, should save a bit more CO_2. In sum, the EU should end up near enough to the Kyoto target of an 8 percent cut to avoid being accused of not practising what it preaches.

The Redesign

Nevertheless a redesign was clearly in order to deal with flaws in the current system and to produce a Third phase of the ETS to take Europe beyond Kyoto, with or without the rest of the world. The Commission had first hoped to unveil its blueprint on the eve of the United Nations climate change conference in Bali in December 2007. In the event, it produced the new design in January 2008. However, the EU did not miss a second UN milestone. On December 12 2008, coinciding with the last day of the UN climate change conference in Poznan in Poland, EU government leaders cobbled together in Brussels agreements on Europe's new climate programme, which the European Parliament endorsed by big majorities on December 17. Some details remained to be formalized in spring 2009, but the basic deals were done.

Passage in less than a year of a reform package – comprising overall emission and renewable energy targets, revision of the ETS, burden sharing between member states, carbon capture rules and subsidies, as well as related proposals on car and fuel emissions – was a remarkable legislative feat. It was also a tribute to President Nicolas Sarkozy's demonic style of chairing the EU during France's presidency of the EU in the second half of 2008. For aficionados of EU policy-making, it took the unusual form of 'first reading' agreements between the Council and Parliament. This required a great deal of preparatory negotiations between key Parliamentary committee MEPs and the French presidency so that the Council and Parliament could both adopt exactly the same amendments at their first plenary votes on the legislation. The political significance of this was to show the unusual willingness of EU legislators – ministers and MEPs alike – to pull together on climate change. How coherent the package is in relation to its aims or how convincing it will be to the outside world is, however, another matter.

As regards emissions caps and allowances, the Commission's aim was to correct the two major flaws of the current system – letting national governments set allocation levels, and giving too many allowances away for free. The Commission succeeded in its first aim, but only partially in the second.

So, replacing the national allocations plans (Naps) that governments have used to 'game the system' to their advantage, after 2012 there will be just one EU-wide cap amounting to a 21 percent reduction in allowances over the 2005–20 period. This will cover all big industries currently coming under the ETS, to which will be added a few extra sectors such as aluminium and the gases of nitrous oxide and perfluorcarbons. In all sectors outside the ETS, such as services, transport, buildings, and agriculture, there will still be national emission ceilings where the Commission proposes differentiation according to the 27 EU states' relative wealth and development. The emission reduction for these non-ETS sectors will be 10 percent. Thus, starting from the new base year of 2005, the overall formula will be: minus 21 percent in the ETS + minus 10 percent in the non-ETS sectors = minus 14 percent in the whole EU economy (see chart below).

The Commission's other main aim was to increase auctioning of allowances to 100 percent by 2013 for the power sector (the biggest user and trader of allowances), and 100 percent for all sectors by 2020. In this the Commission failed, though failure was scarcely surprising given the gathering economic gloom surrounding the December 2008 summit. The general auctioning norm – for all companies outside the power sector and not at risk of carbon leakage (the jargon term for loss of market share or jobs due to carbon constraints) – was set at 20 percent in 2013, rising to 70 percent in 2020, and only hitting 100 percent by 2027. Industries, ranging from Polish generators of coal-fired electricity worried about carbon costs to German machine tool-makers concerned about keeping their number

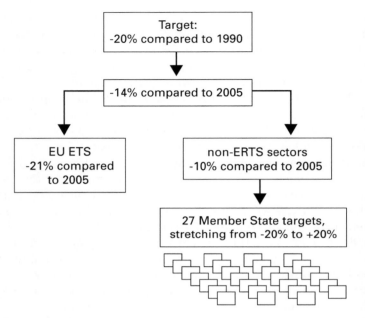

Figure 3: The Commission's New Blueprint for Emission Reduction by
          2020.

Source: European Commission 2008.

one spot in world export markets, won further concessions to
retain free emission allowances (see later for detail). As a result,
the Commission estimates that the rate of auctioning will rise
from today's level of around five percent to nearly 50 percent
by 2013, but will only rise to 60–70 percent by 2020.

This is a pity. Free allowances give rise to windfall profits
for companies that pass on to customers the cost of something
that they, the companies, never paid for in the first place. It
was predictable that energy companies would charge customers,
where they can, the price that given-away allowances fetch in
the ETS; not to do so would incur an opportunity cost. But
such windfall profits are not only politically unpopular at a time
of rising bills for energy users; they can be environmentally
counter-productive, because they delay changes in companies'
behaviour. Such gains insulate managers from the financial

pressure of the ETS to move to low-carbon generation or industrial processes. Regular auctioning would help to ensure that carbon permits figure on companies' books as a real and inescapable production cost and pressure, rather than presenting company treasurers with a financial opportunity that they might or might not exploit. Some individuals in the electricity industry advance the argument that energy companies will perforce spend any windfall profit on new technology or capacity, but this is only a little more plausible than saying lottery winners will perforce spend their windfall on postgraduate courses.

Despite greater centralization of the ETS and the advent of auctioning, there must still be a system of putting national blocks of emission allowances into the hands of national governments. It is governments that will have the right to sell allowances. These blocks of allowances will be distributed among the member states mostly according to relative emissions in the past (though partly also, as detailed later, to give more to poorer states). So there will still be a kind of national allocation. But instead of it being one in which national governments choose the total amount of allowances and then distribute them factory-by-factory around their own companies, it will just be one that awards, on an objective industrial basis, governments their national totals of allowances.

As to the subsequent distribution of emission allowances to companies or factories, a beauty of auctioning is to let the market do the distribution. Under auctioning, bureaucrats no longer have to decide precisely which emissions allowances or cuts to apply to such and such a company, factory or plant. The bidding process – and all national auctions will be open to all EU companies – will do that automatically. But for allowances still given for free, there will still have to be administrative allocation. At the EU level, this means the Commission carrying out the administrative allocation using the technique of benchmarking. It is not clear which benchmarks will be chosen. Taking the historic emissions of a factory or plant as the benchmark for allowances would reward companies that have taken early action to reduce greenhouse gases and therefore have surplus allowances to sell, but would not much penalise dirty technology. More of a spur to action might be to take the best performing technology (in emissions terms) in an industry or sector, and give just enough allowances to cover this benchmark level of industrial perform- ance, but no more. Thus any company or plant falling below this best practice would be penalized by being left short of free allowances. Commission officials claim early phases of the ETS have given them the data, down to the individual factory level, to operate a benchmarking system. But it is not surprising that they would prefer to let the market mechanism of auctioning do the allocating instead.

The rate of auctioning will build up over time. So will the revenues generated by auctioning, even if their increase will be limited by steady annual shrinkage in the volume of ETS allow- ances. This is because of another proposed innovation. Instead of remaining a flat average as it has done in the First and Second ETS phases, the annual cap will decrease along a linear trend line through the 8-year Third phase (see table below).

Table 11: The Carbon Cap Gets Tighter

Year	Millions tonnes of CO_2
2013	1,974
2014	1,937
2015	1,901
2016	1,865
2017	1,829
2018	1,792
2019	1,756
2020	1,720

Source: European Commission 2008

If all sectors in the ETS had to pay for all of their allowances, at a rate of Euros 40 per allowances, auction revenue would rise by 2020 to Euros 75bn a year, or 0.5% of gdp, the Commission estimated in its impact assessment.[2] Partial auctioning, with full payment demanded initially only in the power sector, would produce around Euros 30–50bn a year by 2020, environment commissioner Stavros Dimas estimated when he unveiled the ETS

2   Commission Impact Assessment, SEC (2008) 85/3, pages 10–11 on auctioning.

reforms in January 2008. After the exemptions from auctioning agreed in December 2008, Commission officials revised this revenue estimate downwards to Euros 30bn a year by 2020.

Somewhat surprising is the possible benefit of auctioning to the economy in general. The Commission's conclusion was that if the auction proceeds were fully recycled back into the economy, they would produce less of a drag on the economy than free allocation of allowances. 'Projections indicate that GDP growth, private consumption and employment all could be higher with auctioning than without auctioning for the EU as a whole.'[3]

The logic appears to be that if utilities pass the opportunity cost (at ETS rates) of freely-allocated allowances on to customers, the latter are out of pocket with no useful gain to anyone else. The assumption must be that the windfall gain would just sit idly on company balance sheets (though, equally, it might also be paid out as dividends to shareholders or be reinvested by companies). By contrast, governments can put the auction proceeds back to work in the economy, recycling them either as income tax cuts for households to boost consumer spending, or as reductions in payroll or corporation taxes to boost jobs or investment.

You might therefore have thought governments would have supported the Commission plan for much-increased auctioning. Instead, most governments seemed to prefer immediate relief from pressure by their industrial lobbies, gained by appeasing them with free allowances rather than the more distant promise of greater auction revenue. However, there could be a price to pay for this choice in the UN negotiations. Developing countries have made it clear they want money as well as technology transfer in any global climate arrangement. Politically the easiest way for EU governments to find this money would be to dip into a large pot of auction money. EU governments could commit their taxpayers to make up the shortfall in anticipated auction revenue. But they will not find this a popular promise to make during the recession of 2009, which unfortunately is when the key negotiations for a global accord will take place.

---

3   Annexe to the Impact Assessment, page 62.

The Burden of Burden-Sharing

One of the ways in which the European Commission claims to exercise leadership is in showing the world how richer countries can help poorer ones with the burden of tackling climate change. But there are also sound internal political motives for this, such as avoiding a revolt of the newer, poorer states from central Europe. The latter are regular beneficiaries of the cohesion and structural funds of the EU budget. But the problem was not just that EU climate change policies cost money that these states can afford less than the others. The difficulty was also that the most cost-efficient method of hitting EU-wide targets on policies like renewables would, *unless corrected*, demand a bigger sacrifice from poorer eastern EU states (by exploiting their greater natural potential for expensive alternative energy) than from the richer western ones.

So the Commission proposed three forms of correction:

- In emissions from non-ETS sectors, poorer states would be permitted to expand their emissions by up to 20% (by 20% for Bulgaria, 19% for Romania etc), while richer countries would have to cut their non-ETS emissions by up to 20% (20% for Denmark and Ireland etc). This would, according to the Commission's impact assessment, produce a small increased cost for the Union as a whole – up from 0.58% of gdp (on the least cost scenario) to 0.61% of gdp.
- A slight redistribution of the right to auction ETS allowances – amounting to 10% of the total – from richer to poorer states. Member state governments will hold, and have the right to the revenue from, these auctions which will be open to bidders from anywhere across the EU. So, for example, Latvia will be able to auction off, and keep the money from, slightly more allowances than companies in Latvia would normally use, while Germany would have slightly fewer allowances than companies in Germany would normally use. Such a shift would have no impact on overall cost to the Union, just on income distribution within it.
- National renewable targets (see next chapter for detail) were similarly differentiated. At the two extremes, this gives Romania only a 6.2 percentage point renewable increase in its energy mix, but the UK a 13.7 percentage point increase.

But these three 'corrections' did not satisfy the new member states from central and eastern Europe. They accepted the proposed differentiation in renewable energy targets and non-ETS emission targets; they did not object to a modification making it easier for some richer states to meet their non-ETS targets (see section on credits from third countries). But they insisted on a bigger distribution of auction rights, and demanded a fourth 'correction' in the form of slower phasing in of auctioning for their power sectors.

The emergence of this east-west tension over climate change was not a surprise. In less than 20 years, the East European states had already had to make the unusually sharp energy transition from command-and-control communism to market-based capitalism (even if, as we saw in earlier chapters, the EU energy sector is hardly the freest of markets). Moreover, most of them are at a stage in which they are still intrinsically more interested in economic development than the environment, as well as being more concerned about energy security than climate stability. So they baulked at further costly climate change policies, and they had to be accommodated in a process that was really a microcosm for the wider UN negotiation.

Initially, seven states (Bulgaria, Estonia, Hungary, Latvia, Lithuania, Romania and Slovakia) refused to accept the Commission's way of calculating their national non-ETS emission targets. Even though, in their case, these targets allowed them to increase emissions, the seven states complained that choosing 2005 as the new base year ignored the big emission reduction that the new member states from the east had collectively before 2005. Effectively, they wanted some form of compensation for the post-communist commercial collapse of their polluting heavy industry. For its part, the Commission saw no justification to compensate countries for something that was, first, inevitable and, second, had occurred before they joined the EU in 2004. But the real reason why the Commission refused to budge on its 2005 base year was that this was the first year for which Brussels had solid, verified emission data. Choosing any earlier date would be to put the whole climate change programme on a foundation of sand.

Yet the East Europeans persisted throughout autumn 2008. The Czechs made little trouble, because they had no desire to take over management of an unresolved negotiation during their EU presidency in the first half of 2009. But Poland, which had a very specific demand relating to its coal-dependent electricity, weighed in with characteristic forcefulness. Its prime minister, Donald Tusk, got President Sarkozy to concede at an October 2008 EU summit that any overall deal would be agreed by unanimity (even though, legally, EU environmental issues are settled by majority vote). In fact, it was obvious that a climate deal would be politically unworkable if Eastern Europe felt its views and interests had been overridden by a majority diktat.

In the end, the East Europeans won an increase in the special distribution of auction rights for themselves alone. The December 2008 deal provided for 88 percent of total auctionable allowances to be divided, according to past emissions, among all the 27 states, including the new states. But this last group will also get the remaining 12 percent of allowances to themselves, with the extra 2 percent going to East European states that had already achieved particularly big greenhouse gas reductions.

One final differentiation was agreed in favour of new member states' power sectors. Poland in particular insisted that, with its electricity supply 95 percent dependent on coal, it could not afford full carbon allowance auctioning from 2013. So it and other new member states won the right to phase in auctioning in their power industry, at the rate of 30 percent in 2013 and rising to 100 percent in 2020. This concession was tailored to new East and Central European states by stating that the phased auctioning option was open to states ill-connected to the continental European grid (such as the Baltics) or states at least 30 percent dependent on a single fossil fuel (coal in Poland, gas in Hungary) or states with income per head of only half the EU average (the Balkans).

Will these free allowances lead to windfall profits? Yes, because of the ease with which utilities can usually pass costs, notional as well as real, to customers. To guard against this or at least to find a constructive use for such windfall gains, it was agreed that those states allotting free permits to their electricity companies are to invest in modernising and diversifying their energy system 'for an amount to the extent possible equivalent to the market value of the free allocation'.[4] If it can be enforced, this could be an effective way of narrowing the east-west energy system disparities in the EU. Less sensible was the other anti-windfall profit measure contained in the EU agreement. This gives the new member states the option of preventing the free allowances in their power sectors being traded on the ETS. Poland has indicated it intends to use this option of 'non-tradeability'. This could have the perverse effect of removing the incentive for efficiency for companies that would otherwise be able to sell any free allowances which they no longer need thanks to cleaner or more efficient technology.

Known Unknowns

These imponderables largely arise out of uncertainty about whether there will be a successor regime to Kyoto to which major EU trading partners and rivals would subscribe. This will not be clear until the major United Nations conference in Copenhagen in December 2009 to renegotiate Kyoto.

For its part, the EU has made a unilateral commitment to reduce emissions by 20% by 2020 (from 1990 levels), but would raise this to a 30% cut if other countries take on comparable commitments. The economic slowdown makes any such increased ambition in emission reduction less likely. But if it happened, moving from a 20% to a 30% reduction target need cost the EU very little extra, because the world in which the 30% reduction would take place would be very different from today. It would be a world in which EU companies would have less reason to fear being undercut by non-EU rivals which would themselves be subject to some carbon constraints. It would probably be a world in which other countries would have a demand for carbon allowances and the EU would have less fear of its carbon market price crashing under the weight of carbon credits that could find no other home than Europe's ETS. But, until and unless there is such a wider agreement, the EU knows it will have to cope with several unknowns.

---

4   Elements of the final compromise, Council of the European Union, 17215/08, page 14.

Carbon leakage

The first is the 'carbon leakage' problem – the risk that, in order to be able to compete internationally, industries might decamp, or shift some production, from Europe and set up in countries with no greenhouse gas controls, thereby actually increasing carbon emissions overall. In fact, leakage could occur without EU firms moving facilities or production; it would be sufficient just for EU companies to lose market share and reduce capacity utilisation in Europe.

That EU electricity prices will rise is certain. Though politically inconvenient to say so, Brussels' whole climate change programme depends on prices and carbon costs rising to levels that encourage conservation and reward the generation of low-carbon power that is usually more expensive to produce than electricity from fossil fuels.

How big the rise will be the Commission has found it hard to gauge. One model used by Brussels suggested that the cost increase in electricity generation – related to climate change measures and excluding other factors like gas price rises – could be as high as 33%, while the average cost of electricity (including costs other than generation) would be 19%-26% higher by 2020. But these percentages were probably too high because the starting point baseline was too low. This calculation assumed free allocation of allowances with *no* pass-through of costs, when in reality many companies are already passing on allowance values or costs to their customers. On the opposite assumption that *all* allowances were already being passed on – which did not reflect current reality either – the Commission produced an estimate of a 10%–15% increase by 2020.[5] A reasonable guesstimate might lie somewhere between these two ranges of figures.

The big question for all of Europe's energy-using (or more precisely carbon-producing) industries is whether they can pass the higher power costs on to customers without losing market share to companies with no such costs to shoulder. The answer is clear-cut for Europe's electricity generators. It is most unlikely they could be undercut by non-EU competition. Even if Russia

5   Commission Impact Assessment, SEC (2008) 85/3, page 16

were joined up to the main EU grid, it would be uneconomic for Russia to export electricity (in contrast, of course, to gas) to the EU because of the power losses that occur in long-distance transmission. Hence the decision to make the power sector pay for all its allowances from 2013, and hence the power sector's relative lack of complaint about this.

At the other end of the scale are industries that are heavy users of energy such as producers of primary aluminium or steel (as distinct from scrap re-smelting that requires less electricity) and some basic chemicals. They may find it impossible to pass on much of the extra increase in EU power prices, and impossible to stay in business unless they shift production out of Europe.

Judging which sectors are most at risk will be complex, involving a careful study of the international competition they face. There will, too, be a danger of companies exaggerating the risk in order to get help. Nonetheless, the problem has to be taken seriously. For it is not just a matter of jobs and exports, but also of undermining climate change controls. If the EU were to hit its 20% emission reduction target, and yet give no help to its energy-intensive companies, there would – according to a Commission estimate – be a rise in emissions in other parts of the world equal to 2.5% of total EU emissions.[6] This would be a big leak. Its scale is not, however, surprising. Carbon leakage could very well amount to more than 100% of any carbon reduction made inside the EU, if the market share lost by EU companies goes to companies with more carbon-intensive production processes.

However, counter-measures to deal with carbon leakage would only come into effect, if the December 2009 negotiations in Copenhagen fail to produce a new international accord. Yet, already a year before then, the EU seems to have let itself be panicked by industry lobbies into a drastic definition of the problem and a premature decision on the solution.

At the December 2008 summit, Germany led a successful push for industries at risk of carbon leakage to be given free allowances (albeit up to a benchmark level of the best technology in the sector). Opting for free allowances as a remedy may be premature. While free allowances would probably be a better

---

6   Commission Impact Assessment, SEC (2008) 85/3, page 17.

way of dealing with carbon leakage than imposing EU allowance requirements on makers of imports coming into the EU, they would be worse than negotiating international sectoral agreements on carbon constraints. Such agreements might be feasible for homogenous products like steel and cement.

At the same time, the EU rushed into agreeing on very wide criteria for assessing carbon leakage risk. Sectors or subsectors were deemed to be at risk if the extra direct and indirect costs of auctioning added 5 percent or more to production and if non-EU trade amounted to more than 10 percent of the total size of the EU market. Just in case these two metrics together failed to embrace all candidates deemed worthy of help, a further either/or criterion for risk of carbon leakage was added to bring in sectors with carbon adding 30 percent to production costs or with exposure to non-EU trade of 30 percent. So widely have these criteria been drawn that the Commission estimates they embrace 90 percent of all emissions from EU manufacturing. This is surprising, given that in a regional trading bloc of 27 countries many smaller countries and most smaller companies do virtually all of their trade within the EU.

How many of these free offset allowances for carbon leakage will end up on the market? If a company truly faces a real risk of carbon leakage, there should be no problem about windfall profits, because the company in question would not dare pass on the allowances' cost for fear of losing custom to non-EU rivals. In these circumstances, there is no reason to make these offset allowances 'non-tradeable', because there is not much likelihood they will be offered for sale. But because the terms for offsetting carbon leakage – in the event that there is no global agreement – are potentially so generous, the odds of some abuse are high, with companies either passing the 'opportunity' costs of free allowances to customers or selling free allowances to the market.

Credits from third countries

Another means exists of lowering the cost of the ETS scheme and so minimising the risk of carbon leakage. This is the use of emission credits earned in third countries, another variable or known unknown in the post-2012 equation. These credits were introduced under Kyoto to make targets easier for industrialized countries to meet, and to extend the carbon trading principle of getting maximum emission reduction at minimum cost. But rather like a prescription drug, these are helpful, but need to be used in moderation.

Under Kyoto, the Clean Development Mechanism (CDM) and Joint Implementation (JI) allow industrialized countries to achieve part of their emission reduction targets by investing in emission-saving projects abroad and counting the reductions achieved toward their own targets. CDM covers projects in developing countries, and CDM credits are supposedly given for emission savings additional to those that would have taken place anyway. The JI mechanism covers projects in industrialized countries with a Kyoto target. There is already quite a backlog of unused CDM credits, known as Certified Emission Reductions (CERs), because it has been possible to get retroactive credit for CDM projects back to 2000. In contrast, generation of JI credits, termed Emission Reduction Units (ERUs) could only start in 2008, at the beginning of the Kyoto compliance period of 2008–2012.

These mechanisms were created with a wider carbon market in mind than just the ETS. But, for several reasons, including America's failure to ratify Kyoto, ETS buyers are virtually the only customers for these credits. Brussels is therefore fearful that, in the absence of a Kyoto-2 accord, all these CERs and ERUs will flood into the ETS, swamping the carbon price.

The Commission initially proposed that, in the absence of any further international agreement, after 2012 only those credits agreed to in 2008–2012 could be presented for sale in the ETS. The Commission argued that, because of the CER backlog, EU states could still use these third-country credits to meet around one-third of 2013–2020 emission reduction commitments. However, in the event of a 'satisfactory international agreement' – and therefore of the EU moving to a 30% reduction target – limits on JI/CDM credits would be raised so that these external credits could be used for half that extra EU reduction effort. But not surprisingly EU governments pushed for even greater recourse to outside credits. The eventual agreement allows for external credits to be used for up to 50 percent of the EU-wide emission reductions in the ETS over the period of 2008–20.[7]

Another concession on offsetting, or the use of emission reduction credits earned outside the EU to offset emissions created inside the EU, was agreed at the December 2008 summit to help 11 EU states that might be richer than many of the rest, but also have higher or harder national emission and renewable energy targets than others. Sweden was a good example of the rationale for this extra help. It not only has a demanding national target to reduce emissions by 17 percent in sectors not covered by the ETS (agriculture, transport, services). It also starts from such a high level of renewable energy today (40 percent of total energy use), having almost fully exploited its huge hydro-power potential, that meeting its 2020 renewable energy targets of 49 percent will be extremely difficult. So Sweden – together with Austria, Finland, Denmark, Italy, Spain, Belgium, Luxembourg, Portugal, Ireland, Slovenia and Cyprus – will be allowed to use more outside credits than the other 16 EU states to cover emission reductions outside the ETS.

This last concession brought complaints from environmental groups that the EU was dodging the harder task of cutting emissions at home, and planning to make most reductions abroad. These groups have a point. External credits should be used sparingly. If their import into the ETS were unchecked, they would remove any incentive to make any carbon reductions in Europe. In addition, by lowering the ETS carbon price, they would vastly raise the effective cost of increasing renewables. But if there is an international agreement, these external credits could usefully cushion the impact on the EU of moving from a 20% to a 30% emission reduction target. The EU could meet the latter target with relatively little more emission reduction or impact on its energy system, though it would have to spend more money investing in JI/CDM projects and buying their credits. It might therefore be a mistake to overdose today on credits that would have more effect tomorrow – a similar mistake to

---

7   Questions and Answers on the revised ETS, Commission Memo/08/796, page 9.

wearing an overcoat indoors, so that one does not feel the good of it outdoors.

Transport

There is, finally, a third important unknown in the future of Europe's ETS: how will the US and the rest of the world react to the EU decision to put into the trading scheme emissions from all flights inside, and to and from, the EU?

The legislation the EU has finally decided on is less stringent than some, notably in the European Parliament, had wished for. Aviation's inclusion in the ETS will start in 2012. It will involve initially issuing enough allowances (85% of them free, the rest auctioned) to cover 97% of the 2004–2006 average emitted by aviation in and around Europe, with the amount of allowances decreasing in later years. Nonetheless, this is bound to lead to a confrontation with the US.

The US has promised to mount a legal challenge to what it believes is effectively a tax on fuel; under long-standing international agreements, countries are not allowed to tax aviation fuel. The International Civil Aviation Organization has endorsed the general idea of including aviation in emissions trading schemes, but only if all parties agree – which, in the case of the unilateral EU initiative, they do not. As the legislation was passed in July 2008, the Air Transport Association of America denounced it as 'a tax grab that is not only bad policy but illegal'. President Barack Obama, however generally inclined to participate in an international cap-and-trade scheme, is unlikely to be able to persuade the US airline industry to join in. The US airline sector has long been in poor financial shape, and has strong union and Congressional support for its demands for subsidy and protection.

The case for including aviation emissions in the ETS is that while they account for only 3 percent of total EU greenhouse gases, their dispersion in the higher atmosphere causes disproportionate damage and they have grown by 87% since 1990. Some of this growth is, ironically, the result of the EU's own policy of liberalizing aviation in Europe, a process that created the boom in budget airlines. Congestion in Europe's skies is responsible for extra emissions as planes, stacked up over airports, circle endlessly waiting to land. Unnecessary emissions could be avoided if EU states could agree on a better pooling of their air traffic control systems in the proposed Single European Sky programme.

It is hard not to conclude that inclusion of aviation in the ETS has been driven by its high profile, when at the same time there has been less of a move to include in the ETS the far less visible sector of shipping, which is responsible for just as many emissions as aircraft. The European Commission evidently did not want to chance any disruption to the 90% of Europe's external trade that is carried by ship. But the European Parliament and the Council of Ministers did agree in December 2008 that if the International Maritime Organisation did not agree before the end of 2011 on ship emission targets, the European Commission should propose EU action.

Brussels has not proposed putting road transport emissions into the ETS. This is partly because it had started down a different route of trying to get voluntary improvements from the car industry, partly because there is a complexity of other environmental policies dealing with cars such as green taxation schemes, and partly because the ETS usually deals with 'direct emissions'. That is, the recipients of allowances are the ones directly emitting the CO_2, in other words drivers, in other words millions of individuals. Putting millions of drivers into the ETS would be nonsense. Yet emissions allowances could have been allocated at the level of the car manufacturers' fleets, and these fleet allowances could have been incorporated into the ETS.

However, Europe has decided to deal with car emissions by direct regulation. The Commission initially proposed to oblige car makers to reduce their new car fleets' average emissions down to an average of 130grams of CO_2 per kilometre by 2012. This compares with current average emissions of new EU cars of 160g/km. Complementary action by tyre makers, fuel suppliers and others would contribute another 10g/km of emission savings to meet an overall objective of 120g/km for new cars by 2012.

This proposal provoked a political clash across the Rhine, with Germany rejecting constraints on its heavier or more powerful

Mercedes, BMWs, Porsches, and with France keen to exploit the focus of Renault and Peugeot on smaller cars. "It is hard to argue that heavier, powerful cars with more emissions should have the right to emit more than others", protested French environment minister Jean-Louis Borloo (sounding rather like a Chinese minister complaining about Americans and insisting on human equality in emission levels). Eventually a compromise between President Nicolas Sarkozy and Chancellor Angela Merkel paved the way for agreement at the end of 2008 on EU legislation phasing in the average 120g/km emission limit by 2015 and phasing in penalties on car makers for exceeding this limit. The longer term goal is to get the European car emission average down to 95g/km by 2020, by which time the Commission hopes that the car measures will have contributed one-third of all emission reductions outside the ETS.

The proposed EU standards are tough, particularly when compared to the US. Comparison is easy because emissions are determined by fuel consumption. In 2007, the US Corporate Average Fuel Economy (Café) standards were tightened – for the first time in many years – in order to reach an average of 35 miles per gallon by 2020. If US cars were to meet Europe's 120 g/km proposed standard, they would have to have petrol engines doing 47 mpg or diesel engines doing 52mpg, and not by 2020 but eight years earlier in 2012.

Yet if there were an international emissions scheme that covered the EU and US and their respective car sectors, might not an interesting pattern of allowance trading develop? EU car manufacturers could help pay financially-strapped Detroit to make relatively easy fuel/emission improvements in the US in return for credits that they, the EU car makers, could use to meet their much tighter EU targets. After all, the whole point about global warming is that it does not matter where the emission saving is made, just that it is made.

Conclusion

The biggest single determinant of the success or failure of Europe's climate change programme will be the ETS. This one mechanism covers 40 percent of EU emissions. For all its early trials and errors, the ETS looks to be a workable instrument. Let us hope so – for capping and trading allowances to emit carbon has some important inherent advantages over taxing carbon. It allows maximum emission reduction to be achieved at minimum cost within sectors, within countries, within the EU and internationally. It rewards developing countries' climate control efforts by offering a market for their emission reduction credits. This is a necessary transfer of funds to poor countries, which could one day be supplemented with ETS auction revenue and which would be politically easier for rich countries to carry out than transferring their taxpayers' money.

But the weakness of a cap and trade system is that it cannot provide the absolute carbon price and cost certainty of a straight carbon tax. None of the features of the December 2008 compromise – such as the profligate dispensing of free allowances or the import of more external credits – can confidently be said to impact the future carbon price one way or another. However, together, they are a reminder that Europe's carbon market is very much a political creation, and that the level and stability of the carbon price is vulnerable to politicians' intervention. Tinkering with the ETS should therefore be as infrequent and minimal as possible.

But other factors will have a powerful impact on the carbon price. They include the pace at which low-carbon energy – whether renewable or nuclear – can be developed, and the degree to which energy can be used more efficiently or even not used. Such issues are addressed in the remaining chapters of this book.

CHAPTER 11

MAKING GREEN POWER COMPULSORY

If climate change and CO_2 emissions were the sole goal of energy policy, and the renewable energy sector were a mature and well functioning market, then a single CO_2-based target would be appropriate — but this situation is a long way off.

European Commission impact assessment, 2006.

The renewable energy target serves more than just reducing greenhouses gases.

European Renewable Energy Council, 2007.

The revival and development beyond all recognition of some of mankind's most ancient forms of energy, such as wind and water power, has provoked a very modern debate in Europe about policy goals and costs. The debate suddenly acquired a real edge to it after EU leaders surprised many, including perhaps themselves, by agreeing at their March 2007 summit that renewable energy must rise as a share of total energy consumption to 20 percent by 2020. Some leaders, it is said, misunderstood the '20 percent' just to be a share of electricity, a far lesser goal. At all events, they may all have rued this decision when ten months later the Commission handed them its proposals for the binding national renewable energy targets necessary to deliver the EU commitment.

Within a decade, renewable energy has gone from a nice-to-have to a must-have component of Europe's energy mix. It is the only sector (along with its sub-sector, biofuels, see next chapter) to be singled out for such special treatment by the politicians. This special treatment started in 1997, when the Commission proposed 'an indicative objective' for renewable energy to reach 12 percent of energy consumption.[1] At the time, the EU executive was of the view 'that an indicative target is a good policy tool, giving a clear political signal and impetus for action'. This was to prove optimistic.

Four years later, the EU passed a directive setting a target, indicative again, of 21 percent for the share of electricity to be generated renewably by 2010. By 2007 the Commission judged that the EU was on track to reach a 19 percent share of renewable electricity by 2010, only a couple of points off the target. Outside of electricity, however, renewable energy had made little inroad.

By 2007, abandoning its earlier benign view of the effectiveness of voluntary indicative targets, the Commission was complaining that 'the absence of legally binding targets for renewable energies at the EU level, the relatively weak EU regulatory framework for the use of renewables in the transport sector, and the complete absence of a legal framework in the heating and cooling sector, means that progress is to a large extent the result of the efforts of a few committed member states.'[2] In wind power, the EU's three leaders are Germany, Spain and Denmark, far ahead of the rest. Finland and Sweden are the biggest burners of biomass for electricity. The photovoltaic sector is dominated by Germany with 86 percent of current installed PV capacity in the EU, a bizarre ratio reflecting subsidy rather than sunshine. So the Commission concluded that only mandatory targets could produce a more even performance for renewables across sectors and across countries.

The main rationale for promoting renewables is to reduce carbon emissions. Hitting the 20 percent target would save 600–900m tonnes in CO_2 emissions a year, the Commission claimed.[3] But there are other forms of low-carbon energy, notably nuclear, and cheaper ways of cutting emissions such as energy efficiency and demand reduction measures. So promoters of renewables also vaunt their other merits in providing energy security and employment. The EU will also save money on importing fossil fuels, as much as 200–300m tonnes a year according to the

1   'Energy for the Future: Renewable Sources of Energy', Commission White Paper for a Community Strategy and Action Plan, COM (97) 599, 1997, p.10.
2   Renewable Energy Road Map, COM (2006) 848, p. 5.
3   Commission Memo/08/33, p. 3.

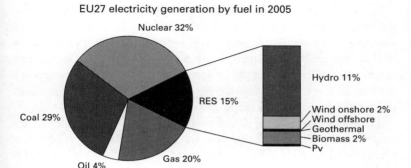

Figure 4: Renewables' Place in the Generation Mix

Source: European Commission document, SEC(2008)57, p.18

Commission, and give itself greater diversity of energy sources, strengthening Europe's resilience in the event of external shocks such as oil interruptions. Another gain would be the boost to Europe's renewable industry that already has a turnover of Euros 30bn a year, employs some 350,000 people, and provides alternative custom for Europe's farmers and foresters.

But there is a price tag on going green. This can be calculated as the total cost of renewable generation minus whatever conventional fossil fuels might cost in the future. The higher the oil price (to which the gas price is mostly linked), the lower the real net cost of renewable. The cost of renewable generating equipment might also vary, but not so dramatically as the oil price, and it could drop. So, at a $48 oil price the additional annual cost of moving towards the 20 percent renewable target would be $18bn, but this would sink to $10.6bn a year if the oil price rose to $78 per barrel.[4] It must be said this Commission cost estimate for the whole EU looks understated, if there is any accuracy to the UK's 2008 forecast for its own renewable costs by 2020. This forecasts an extra £5–6bn a year by 2020, on the assumption that oil would be around $70 a barrel then.[5]

---

4   Renewable Energy Road Map, COM (2006) 848, p.16.
5   Department for Business, Enterprise and Regulatory Reform, UK renewable energy consultation, June 2008.

This extra price is worth paying if, as the citation from the European Renewable Energy Council at the start of this chapter suggests, a value is put on energy security and employment as well as on the reduction of emissions. Reaching a 20 percent renewable share in Europe's energy mix is not strictly necessary for the EU to hit its over-arching goal of a 20 percent emission reduction, but it could bring these other benefits.

However, there is a risk that meeting the renewable target could, at the margin, hamper progress towards the greenhouse gas reduction goal. This is because of its effect on the ETS carbon price, which is, or should be, a neutral driver pushing forward all low-carbon technologies from nuclear power and carbon capture and storage (CCS) to renewables. The paradox is that if any of these low-carbon technologies is pushed artificially hard – through non-market mechanisms, such as targets, rules or government fiat – the effect will be to depress the carbon price simply by pushing demand for carbon allowances on the ETS artificially low.

Commission economists have run projections showing that, everything else being equal, meeting the twin 20 percent emission and renewables goals simultaneously would produce a carbon price of Euros 39 a tonne of CO_2 by 2020, compared to Euros 49 a tonne if the greenhouse gas target alone were allowed to drive renewables.

Thus, it is possible that the emissions target might not be met if the incentives to develop nuclear and/or CCS were sufficiently undermined by a weaker carbon price. The extent of any undermining would depend on how much carbon prices actually prove to be the deciding factor in nuclear or CCS investment rather than regulatory obstacles and planning delays. And, if the carbon price did prove key, it could be supported by withdrawing some carbon allowances from the market, though such intervention might damage belief in the market's integrity.

In theory, the minimum 10 percent biofuel target (see next chapter) could also weaken the ETS carbon market price, because it is another non-market mechanism being used to boost low-carbon energy. In practice, it will have less effect on the carbon price. This is because the transport sector, where most of the petrol and diesel carbon emissions displaced by biofuels will occur, is not covered by the ETS, though the process of manufacturing fossil fuels in oil refineries is. Yet the biofuel target does introduce artificiality. For without such a target, biofuels would, as one of the most expensive renewables, be one of the last to be developed. With the target, they may displace some other renewables.

Of course, it is true that the EU carbon market could have been bent more out of shape if EU leaders had followed the European Parliament, which had originally wanted a renewables target of 25 percent of final energy demand by 2020 (and an indicative 40 percent target for 2050). It is also the case that aiming now at a 20 percent renewable target might prove a useful building block if the EU subsequently went for a higher emission cut. For the EU has clearly said that while its 20 percent emission cut (from 1990 levels) is unconditional, irrespective of what the rest of the world does, it would move to a 30 percent cut if this were matched internationally.

National targets

However, EU leaders only did part of the job when they agreed at their March 2007 summit to the 20 percent average target for the Union. The trickier part was to break this down into binding national targets. The leaders gave the following guidance in their summit conclusions:[6]

> Differentiated national overall targets should be derived with member states' full involvement with due regard to a fair and adequate allocation taking account of different national starting points and potentials, including the existing level of renewable energies and energy mix, and, subject to meeting the minimum biofuels target in each member states, leaving it to member states to decide on national targets for each specific sector of renewable energies (electricity, heating, cooling, biofuels).

But this left the Commission with guidelines that potentially

---

6   See http://www.consilium.europa.eu/ueDocs/cms_Data/docs/pressData/en/ec/93135.pdf

conflict (starting point versus potential, for instance). So Brussels officials looked at various options:

- One was to repeat the classical modelling exercise that Brussels had used to produce the indicative national targets contained in the 2001 directive. The basic technique here was to increase the marginal cost of conventional energies and see at what point on their cost curve renewable energy could begin to compete with them. However, such an approach produced very different results for different countries, in particular high targets for the central and eastern European countries that were not part of the EU in 2001. Modelling brought out these new member states' high renewable potential – they had generally done little to 'go green', yet had considerable biomass to do so – and set them correspondingly high targets. Another difference with 2001 was, of course, that this time the targets were binding, and therefore as one official said, 'member states will always try to out-model us, or quibble with our assumptions, if they don't like the result'.
- Asking every member state to make by 2020 the same 11.5 percent point increase from their actual 2005 renewable share. But it was felt this would be unfair on those states that had done a lot already or had little extra potential to do more. Several countries would fall into both these categories.
- Facing such difficulties about national targets, the Commission even thought briefly of putting targets and constraints on companies rather than governments. So all companies would have a target or supply obligation, such as every oil company would have to make 10 percent of all fuel sold biofuel. But it was quickly realized that while such an approach could be applied to big operators (electricity and oil companies), it would be impossible to apply to the individualised sectors of heating and cooling.

Therefore, the Commission decided to combine some of the options in a way that would respond to the March 2007 summit's call for fairness. The overall goal was to raise renewables' share in final energy demand from 8.5 percent in 2005 to 20 percent by 2020. Half this 11.5 percentage point gap would be closed by an equal increase to every state's renewable target share, and the other half with increases varied to take account of relative gdp and, to a small extent, states' green energy starting point and potential. At one extreme, this gave Romania only a 6.2 percent point renewable increase in its energy mix. At the other was the UK with a 13.7 percent point increase; Britain lags far behind in renewable development, ahead of only Malta and Luxembourg, but as a windy island has obvious wind and tide power potential. The Commission judged the balance right. The

Table 12: National Renewable Targets

	Share of energy from renewable sources in final consumption of energy, 2005 as a percentage	*Target for share of energy from renewable sources in final consumption of energy, 2020 as a percentage*
Belgium	2.2	13
Bulgaria	9.4	16
The Czech Republic	6.1	13
Denmark	17.0	30
Germany	5.8	18
Estonia	18.0	25
Ireland	3.1	16
Greece	6.9	18
Spain	8.7	20
France	10.3	23
Italy	5.2	17
Cyprus	2.9	13
Latvia	34.9	42
Lithuania	15.0	23
Luxembourg	0.9	11
Hungary	4.3	13
Malta	0.0	10
The Netherlands	2.4	14
Austria	23.3	34
Poland	7.2	15
Portugal	20.5	31
Romania	17.8	24
Slovenia	16.0	25
The Slovak Republic	6.7	14
Finland	28.5	38
Sweden	39.8	49
United Kingdom	1.3	15

Source: European Commission 2008

targets (see table below) that it proposed in January 2008 were almost exactly what the Council of Ministers and European Parliament agreed in December 2008

Trade in renewables

Differential targets may produce fairness. But since some states have target increases above – and others below – their renewable potential, there is a desire, indeed a need, to correct this imbalance through some cross-border trading of renewable energy or certificates of renewable energy. And satisfying this desire has proved very difficult because of national subsidy schemes for renewables.

Brussels has found designing trade in renewable certificates much harder than trade in emission allowances. It has had to steer around 27 different national renewable schemes, while only two countries – Denmark and the UK – had launched their own emissions schemes before the ETS was created. Moreover, while Denmark and the UK did not mind seeing their emissions schemes being subsumed into an EU-wide scheme, there is deep attachment by many member states to their existing national renewable schemes. These mainly divide into two categories:

- 18 member states operate feed-in tariffs (a guaranteed full price) or premiums (a bonus on top of the electricity market price) paid to producers for the renewable power they feed into the grid. The premium system allows the market more of a role than the feed-in tariffs, but they both provide long-term certainty that is evidently valued by investors. The three biggest developers of renewables – Germany, Spain and Denmark – use feed-in tariffs. These tariffs and premiums can also be set at different levels to stimulate more distantly commercial technology such as solar PV as well as relatively low-cost onshore wind power.
- Seven member states impose quota obligations on suppliers to source a certain percentage of their electricity from renewable sources. This is usually facilitated by a tradeable green certificate (TGC) scheme. So the renewable energy

generators, who are the object of the subsidy, sell their green power for whatever they can get in the electricity market, but they also can sell accompanying 'green certificates' to suppliers who need the certificates to show they have fulfilled their quota obligation. This system gives renewable producers less certainty, because their income depends on two fluctuating values, the market price of green certificates and the market price of electricity.

Because of the element of trade built into it, the second scheme can obviously be more easily adapted to cross-border trade. Indeed there already is a cross-border market in the transfer or trade in 'guarantees of origin' (GoOs), produced by various issuing bodies to certify that units of electricity have been renewably produced. By contrast, feed-in tariff systems were not designed for their benefits to be traded separately from physical delivery of the electricity. Instead, feed-in tariffs are intended to reward electricity that is actually fed in to the grid of the country offering the tariff.

However, cross-border trade in renewable energy is needed to build up economies of scale across Europe, to move investment where it will produce the best return, and to help countries meet renewables targets that do not match their potential. Because physical flows of electrons cannot be precisely tracked across the multiple borders in the EU, the only way for a pan-European renewable market to operate is on the basis of virtual trading in guarantees of origin, unhooked from the limitations of physical delivery. And this was what, in principle, the Commission proposed. So, as an example, a Greek producer of solar power would be able to present its guarantee of origin in Germany in order to get a much higher solar feed-in tariff (perforce, because Germany has less sun than Greece), even though the Greek solar power might never reach the German grid. Likewise, German renewable energy producers might present their guarantees of origin for sale on the UK Renewable Obligation Certificate market, even though their power would never actually cross the Channel and still be sold on the German market.

This prospect stirred fears in member states, particularly those with feed-in systems, that governments would simply lose control. As the German environment and Spanish industry minister put it in a joint letter of complaint to the Commission (just before the EU executive unveiled its plans on 23 January 2008), 'if member states have to achieve a national target, they need to have the means in their hands and they must not lose these means through an EU-wide scheme.' There were worries about uncontrolled inflows of green power (in the form of GoOs being presented for feed-in payments) that might push a country unnecessarily over its renewable target and at an exorbitant cost. Equally, there were concerns about outflows of green power from countries that would then undershoot their targets.

The major reason, of course, for such inflows and outflows would be to exploit the differences in feed-in tariff or premium levels between various EU states. The effect of uncontrolled trade over time would be to reduce these differences, and to make it hard for governments to set their own tariff levels in the future. This prospect of de facto harmonization has been resisted by 'feed-in' countries, and the renewable energy industry itself, as fiercely as any formal attempt by Brussels to propose an EU-wide support scheme.

In other sectors of the European economy, the Commission would regard a multiplicity of state aids as dangerously distorting and would use its autonomous powers to rein in these state aids or, at a minimum, harmonize them. It has had to take a different attitude to renewable energy. State aid is accepted as essential because renewable energy is considered an unqualified public good, and because Brussels has no comparable EU money to promote it (see section on carbon capture and storage in Chapter 14).

Some Commission officials would like to harmonize national support schemes. They realize delay merely stores up trouble for the future. Indeed the 2001 renewable directive seemed to offer a chance to end the fragmentation of support schemes. It required the Commission to report in 2005 on the cost effectiveness of the various national support systems, on whether to harmonize them and if so on what model.

But when 2005 came around, the Commission dodged the issue. It said the track record of feed-in and quota obligations were too short to make a proper comparison and gave itself another two years to answer the question. In 2007 the Commission came off the fence slightly. Its report then found that '*well-adapted* [original italics] feed-in tariff regimes are generally the most efficient and effective support schemes for promoting renewable electricity', as the chart below indicates. Yet the Commission went on to say that 'while harmonization of support schemes remains a long term goal on economic efficiency, single market and state aid grounds, harmonization in the short term is not appropriate.'[7]

The figure shows how spectacularly effective feed-in tariffs have been over recent years in Germany, Spain and Denmark.

The Commission's dilemma is that the system (quota obligations) most apt architecturally for the whole EU appears to be

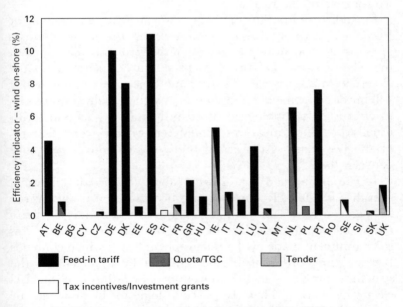

Figure 5: Effectiveness of Renewable Subsidies, 1998–2006

Source: European Commission document, SEC(2008)57, page 26. The effectiveness indicator is the ratio of increased electricity to additional realisable potential over the same period.

---

7   The support of electricity from renewable energy sources, COM (2008) 19, p. 17.

less effective in actually increasing green power than the system (feed-in tariffs) less suited to the EU scale.

Both systems can be prone to over-paying companies that thereby reap windfall profits. Feed-in tariffs need fine tuning, usually downward adjustment, to take account of technical progress as technology matures, but most tariffs have this degressivity, or downward tapering, built in. In the quota obligation system, it is the most expensive technologies that set the marginal cost of meeting the quota obligation and determine the price of tradable green certificates. So anyone operating more cost-efficient technology, typically onshore wind power, will benefit more. But feed-in tariffs seem to score better on effectiveness, in attracting investment, because they provide financial certainty irrespective of the market.

Such divorce from the market is a weakness from a national or European viewpoint. It is therefore welcome that some countries are moving away from pure feed-in tariffs (which totally supplant the electricity market price) to premiums (which top up the electricity market price). At the same time, the UK has said it will introduce some differentiation in its quota obligation system to encourage technological diversity in the way feed-in tariffs usually do. Such measures should remove the worst features of the two main systems, and represent a slight convergence between them.[8]

It can be argued that fussing about distorted renewable subsidies is relatively unimportant, because even if the overall 20 percent renewable target is met, this will only increase renewables to about a third of the EU electricity market. This is an argument made by the renewable industry, which contends Brussels' first order of business should be to tackle all the structural problems in the conventional two-thirds of the EU electricity market. There is a certain logic to this sequence of events. Removing discrimination in the internal energy market ought to make it easier for renewables to get on the grid, though there are technical, planning and financial issues that can also make that hard.

But the difficulty of general internal market reform should

---

8   Ibid. p. 15.

not be allowed to become the pretext for indefinitely delaying the creation of a coherent and consistent EU renewable support programme. The Commission finally said as much in the January 2008 launch of its renewable energy policy. 'When the single electricity market becomes competitive and new entrants producing renewable electricity can participate on a level playing field, certain design features of renewable electricity support schemes will have to be reviewed.'[9]

Restrictive trade practices

For the foreseeable future, however, it looks as though trade in green energy will be very restricted, out of deference to the big feed-in tariff countries and the renewable industry. The trade will probably be controlled by governments, as it was between members of Comecon, the Soviet bloc economic organization, which is not a great advertisement for any system.

Under pressure from the start to restrict trade, the Commission originally proposed that governments would be able to set up a system of prior authorization for the transfer in and out of their territory of these GoOs if they were concerned about maintaining their support schemes or hitting future renewable targets. Prior authorization would give the states the ability to vet, and the right to veto, green certificate transactions. Nor would renewable generators become free to go subsidy shopping around Europe. Out would go the current restriction tying a generator to the support scheme of the member state in which it is physically located. But in would come a new 'lock-in' restriction tying a generator to whichever member state it first presents a GoO; this could be another member state, though it would most likely be the generator's home state.

Nonetheless, this was still too 'free trade' for heavyweight renewable states such as Germany, for the renewable industry and, crucially, for Claude Turmes, the Luxembourg Green MEP who was the European parliament's rapporteur on the renewable directive. Mr Turmes' suspicion of the Commission plan was

---

9   Ibid. p. 13.

increased by the backing it got from the Eurelectric organization of big generators and the European Federation of Energy
Traders. 'Creating an EU wide renewables certificate market
is not the way forward', Mr Turmes wrote in his report. 'It
would undermine the existing national support schemes, but also
potentially generate Euros 30bn in windfall profits for traders
and generators', on the ground it would favour technology with
the lowest marginal costs like onshore wind to the exclusion of
other more exotic technologies. This, he concluded, would far
exceed the potential Euros 8bn a year saving by 2020 that the
Commission had calculated could be gained by having EU-wide
trading in green power certificates.[10]

By mid-summer 2008, most proponents and opponents of
certificate trading had become bogged down in what one Commission official described as '1914–18 trench warfare'. It was at
that point the UK, Germany and Poland got together to suggest
a compromise. This would allow some trading across borders
and even outside the EU. But it would crucially leave governments in charge of any trade of renewable energy, related to
fulfilment of their national targets, which could be exchanged
on the basis of official statistics.

This proposal preserves some of the Commission's plan for
'virtual' renewable energy trade, but puts it all under governments' control. It could take the form of statistical swaps
between member states (which would buy and sell percentage
points of green power), or two or more member states combin-
ing targets or support schemes, or deals between a couple of
member states whereby a renewable project would be built in the
first state but some or all of the energy would count towards the
goal of the second state. 'We want a single market in renewable
energy, but not, at this stage, a single market in renewable energy
finance', commented a UK official.[11] The UK–German–Polish
proposal was the basis for the renewable trading system agreed
in December 2008.

In conclusion, the problem is not that the EU – for

---

10 Claude Turmes, Environment Committee report on the renewable
    energy directive, 26 September 2008
11 Author Interview 2008

understandable political reasons of solidarity and equity – settled on a system of differentiated national targets. The pity, rather, is that having come up with a system that requires cross-border trade, the EU then did its best to frustrate that trade. Significantly, in its otherwise gently-worded report in 2008 on EU energy policy, the International Energy Agency was sharply critical of EU restraints on renewable energy trading.

Another criticism of the renewable energy targets is that they are impossibly high, certainly for a country such as the UK with a record of target failure; failing to achieve them or achieving them at excessive cost will discredit the whole programme. This is the charge made, among others, by Dieter Helm who has suggested various ways of softening the target (partly by redefining renewable as low carbon to embrace CCS or even nuclear and partly by prolonging the deadline beyond 2020).[12]

Of course there is a psychological point at which, if the bar is set too high, you don't even bother to try to jump. And at 20 percent the renewable bar might prove so high as to be incredible to the wide number of market players needed to create a broad renewable energy base. Yet so-called stretch targets can be good. If targets stretch the abilities or efforts of people, companies or states in a good cause, they are beneficial even if they are not met. However, while ability and effort are not finite and can and should be stretched, natural resources are finite. This is why the one renewable goal under real attack has been the sub-target set for biofuels.

---

12 Renewables – time for a rethink? June 2008. www.dieterhelm. com.

CHAPTER 12

PUTTING TROUBLE IN YOUR TANK

At the rate at which the European Union and its member states are supporting the production of ethanol, they could have gone to the world market and bought twice as much energy in the form of petrol for slightly less money.

2007 report by the Global Subsidies Initiative.

Not all biofuels are equal – there should be no favouring of EU production of biofuels with a weak carbon saving performance if we can import cheaper, cleaner biofuels.

European trade commissioner Peter Mandelson, July 2007.

Despite biofuels being cast as the culprit for pressure on world food prices, there is a case for *some* further increase in the use of biofuels in Europe.

Road transport accounts for nearly one third of Europe's total energy use. Around 98 percent of road transport is fossil-fuelled. Most of the future growth in Europe's CO_2 emissions will come from transport. And biofuels are the only cleaner alternative road transport fuel on the horizon. Moreover, replacing some of Europe's imported oil with home-grown fuel improves energy security, and in a small way moderates the rise in oil prices. According to the International Energy Agency, 'biofuels have become a substantial part of faltering non-Opec supply growth, contributing around 50 percent of incremental supply in the 2008–13 period.'[1]

So in March 2007 European Union leaders decided biofuels should, in principle, account for at least 10 percent of all transport fuel in all 27 states in the Union by 2020. In the January 2008 draft legislation to implement this goal, the Commission proposed the 10 percent minimum should be of 'renewable

---

1   IEA Medium Oil Market Report, 1 July 2008

energy', not just biofuels. This redefinition was retained in the December 2008 legislative agreement, which made clear the 10 per cent renewable energy minimum should be of the EU's total fuel consumption in all forms of transport. The final deal gives a preference to the development of so-called second generation biofuels – such as fuels made from residue, waste and woody biomass – which unlike crop-based first generation biofuels do not compete with food or feed production. So second generation biofuels will get a double credit towards the 10 percent target, while renewable electricity powering electric cars will be counted at 2.5 times their input towards the target. Green electricity powering trains can also count towards the target, but only once as with all first generation biofuels.

But before delving into the controversy behind this shift in emphasis, it is important to establish why a mandatory across-the-board minimum was felt to be necessary in the first place. It is not just that it suits European farmers and those EU states with a big farm lobby as a continuation of the Common Agricultural Policy by other means. There is another reason. If biofuels were bundled in with other forms of renewable energy, and left without a specific target, many people and governments in the EU would think it more environmentally or economically rational to focus on wind or solar power or even other uses of biomass.

For if you wanted to use biomass – crops, wood and waste – to get maximum reduction in greenhouse gases you would use it for electricity, and if you wanted to turn biomass into energy most efficiently, you would use it for heating. So, if there were no compulsion to develop biofuels, nothing would be done to clean up Europe's vehicle emissions. (The only profitable form of biofuels developed so far remains alcoholic spirits for human consumption. 'Biofuels are basically booze', a vice-president of the ExxonMobil oil company recently told a conference, 'and we don't do booze.')

For those who savour trade-offs and policy dilemmas in energy policy, biofuels are a gem. The biofuel industry will compete with the food sector for agricultural crops. It may, while helping to combat global warming and to clean up the atmosphere, also damage the terrestrial environment by encouraging monoculture of energy crops and reducing bio-diversity. As a relatively clean home grown form of energy, biofuels would appear to appear to serve the cause of both energy security and climate stability. But there could be friction between these two goals, especially if, because of its protectionist biofuel lobby, the EU were to aim at biofuel self-sufficiency by growing biofuels that only marginally reduce carbon and by shutting out imports with a far higher 'carbon-saving' capacity. That in turn could lead to conflict with many developing countries that see in biofuels a valuable new export. The climate could also suffer if Europe were to import biofuels heedless of whether these had been produced on land cleared of rain forest; for halting tropical deforestation is by far the most effective way of slowing the rise in carbon emissions.

Aware of some of these pitfalls, EU leaders attached some conditions to their March 2007 summit's endorsement of the 10 percent biofuel target for 2020. They said it should be introduced 'in a cost-efficient way', and added that 'the binding character of this target is appropriate subject to production being sustainable, and second generation biofuels being commercially available'. But, perhaps unwisely, the Commission did not take this too seriously. It entered a statement into the minutes of the March 2007 summit that it 'does not consider the binding nature of the target should be deferred until second generation biofuels become commercially available'. Subsequently the Commission official in charge of renewable and biofuel policy told a conference that the rider about second generation development should not be regarded as 'absolute conditionality'. However, as we shall see, the European Parliament has taken this condition rather more to heart.

Although used in Europe during periods of war or excess agricultural production, biofuels only became the object of serious scientific research and political attention after the first oil shock of the early 1970s, and of industrial production since the early 1990s. The first policy measures to benefit biofuels were not specific to the industry at all – the CAP was reformed to divert agricultural surpluses to industrial uses. As part of its deal with the US concluding the Uruguay round of world trade negotiations, the EU instituted a scheme whereby farmers had to set aside a portion of their arable land, on which they could

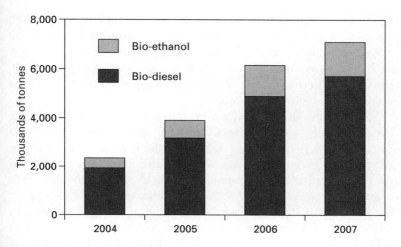

Figure 6: Biofuel Production in the EU

Source: Sources: European Bio-diesel Board, European Bio-ethanol Fuel
Association.

grow non-food crops, such as oilseed rape for biodiesel. The
Commission reported in 2006 that more than 95 percent of the
'non-food set-aside areas' had been used for energy crops.[2]

A 2003 directive on the voluntary promotion of biofuels set a
non-binding target for a 2 percent biofuel share of the EU road
fuel market by 2005, and 5.75 percent by 2010.[3] But by 2005 the
actual biofuel share was only 1 percent, and it became clear that
voluntary means alone would be insufficient to meet the 2010
goal, despite the existence of sizeable fiscal incentives.

Economic costs

The biggest financial prop for biofuels has been exemptions
from, or reduced rates of, excise duty on fuel. There is no EU-
wide exemption from excise duty for biofuels – partly because
there is no EU excise duty or common EU-wide level of national

---

2   COM (2006) 500, Report from the Commission to the Council on
the review of the energy crops scheme, p. 7.
3   EU directive, see 2003/30/EC.

excise duty on fuel. But there is EU legislation allowing member states to give biofuels exemptions from their standard rate of excise on fuel. In the 1990s these fiscal advantages were limited to biofuels produced in pilot plants, but since the passage of the 2003 Energy Taxation Directive they can, and do, cover commercial biofuel production.

As of mid-2007, 16 member states were offering such tax breaks to their biofuels sectors. Since the aim is to enable biofuels to compete on equal terms with fossil fuels in the marketplace, the tax break is supposed to cover no more than the gap between oil prices and biofuel production costs. Nonetheless, these tax exemptions constitute by far the largest part of financial support for biofuels in the EU. By one calculation, they amounted in 2006 to over Euros 900m of the Euros 1.3bn total financial support that went to bioethanol in Europe, and Euros 2.1bn of the total Euros 2.4bn that went to biodiesel in Europe.[4]

However, the burden of support will spread more widely to consumers as well as taxpayers, as the compulsory minimum market share for biofuels comes into effect. Some such quotas are already here on a national level. By 2008 nine member states had already, on their own initiative, imposed mandatory biofuel market shares or blending targets on themselves. They included, ironically, the UK, a country that lags behind almost all others in its take-up of biofuels. The UK introduced its 'renewable fuel obligation' on April 1 2008, only to respond to growing public disquiet about the biofuel impact on food prices by announcing a couple of days later a review of the policy by Ed Gallagher, chairman of the UK Renewable Fuel Agency.

To clean up conventional road fuels, the EU also agreed at the end of 2008 on a revision of its 1998 Fuel Quality Directive. As well as raising the amount of biofuel that can be blended with petrol from 5 percent to 10 percent, the revision would require a 6 percent reduction in greenhouse gas emissions per energy unit of fossil fuel by 2020.

The cost of supporting biofuels is bound to increase in the

---

4   'Biofuels – At What cost? Government Support for Ethanol and Biodiesel in the European Union – 2007 Update' Global Subsidies Initiative, Geneva, October 2007.

future. The biggest element in that support – exemption from excise duty – would evaporate if biofuel production costs fell below that of oil. Even at the very high oil price of mid-2008, this is unlikely to happen, partly because some fossil fuel is needed to make biofuels. As the Organization for Economic Cooperation and Development (OECD) has pointed out, 'higher oil prices will both raise the production cost of biofuels (as fossil fuels are an important input in the production process) and exert upward pressure on agricultural commodity prices as a result of the increased demand for them.'[5] So by expanding biofuel production, the positive link between oil prices and biofuel costs might awkwardly get stronger, not weaker. If the cost of biofuels moves in the same direction as oil prices, the biofuels would be unlikely to reduce transport prices.

If all that mattered was producing biofuels, the EU could do a great deal. According to the Commission's Biofuel Research Advisory Council, 'in 2030, EU biomass would hold the technical potential to cover between 27 percent and 48 percent of our road transport fuel needs, *if all biomass would be dedicated to biofuel production*'(emphasis added).[6] But, in the absence of war or total and prolonged interruption in oil imports, devoting all biomass to making biofuel is a quite unrealistic proposition. So the advisory council settled for a quarter share of EU road transport fuel needs being covered by biofuels in 2030 as 'realistic', half from domestic production and half from imports.

Even a quarter-share could be fanciful, however, according to a report done for the OECD. It believes the economics of biofuels will remain unfavourable. 'Although there is scope for production costs for biofuel feedstocks to decline as a result of improvements in yields, it is not clear that such improvements will be enough to compensate for rising prices due to production factors and the combined pressures on prices of rising demand for food, feed and biofuels. Increasing competition with biomass feedstocks – woody material as well as agricultural products – is

5 OECD report, 'Biofuels: Is The Cure Worse Than The Diseases?', 12 September 2007, p. 5.

6 Final report by the Biofuel Research Advisory Council, Office for Official Publications of the European Communities, 2006, p.18.

actually pushing feedstock prices and production costs up. Higher oil prices will have the effect of increasing biofuel production costs while simultaneously making fossil fuel alternatives such as tar sands and coal-to-liquids increasingly competitive.'[7] All of these are factors that would threaten the economics of all but the most competitive biofuels such as Brazilian ethanol.

Environmental costs

The most obvious tensions in promoting biofuels are the risks to food production and the environment. Controversy rages over biofuels' share of the blame for higher world food prices. The US administration and the European Commission put this share as low as 3 percent, but an internal World Bank report was reported to blame biofuels for 75 percent of the 140 percent rise in the price of a basket of food commodities over the period of 2002–8.[8] For its part, the UK's Gallagher review concluded in July 2008, 'the demand for biofuels contributes to raise prices for some commodities, notably for oil seeds, but that the scale of their effects is complex and uncertain to model.'

The Commission claims to be relatively confident that, at least in the short to medium term, the strains on EU crop resources would be manageable, provided that the EU lets in adequate imports and makes progress, over the longer term, on second-generation biofuels made out of wood and cellulose that would not compete with food. EU production of ethanol is relatively modest, using less than 1 percent of the Union's cereal and sugar beet harvests. But any further surge in biodiesel production in Europe could put serious pressure on rapeseed oil output, of which 60 percent already goes to biodiesel. The scientific committee of the European Environment Agency, an EU body, gave in April 2008 its view that the proposed 10 percent target was 'over-ambitious', carried too many environmental risks, and should be suspended pending further research and replacement by 'a more moderate long-term target, if sustainability cannot be

---

7   OECD report cited above, p. 6.
8   Reported in *The Guardian* newspaper, 4 July 2008.

guaranteed'.[9] Increasing concern has also been expressed about the 'displacement effect' of increased cultivation of biofuels in Europe, leading to more land being cleared in developing countries for the food that Europeans would no longer be growing. Moreover, the food industry is not biofuels' only competitor for the produce of Europe's fields and forests. Outside the energy field, there are other industrial users of biomass, especially chemical companies that draw many substances from agriculture and forestry, and the packaging and construction sectors that use a lot of wood products. The governments of Austria, Belgium, Finland, France, Germany and Luxembourg – all with forestry interests – made a joint appeal in December 2007 for the EU not to let its drive for biofuels short change these other industries of renewable raw material.

The environmental calculation has to weigh what a given biofuel process does for the atmosphere and the land. Specifically, can it 'save' enough greenhouse gases, compared to conventional petrol and diesel, to justify the extra strain it might put on the land? OECD studies claim only three current technologies meet this test: Brazil's sugarcane-to-ethanol process; ethanol produced as a by-product of cellulose output as in Sweden and Switzerland; and manufacture of biodiesel from animal fats and used cooking oil (requiring little or no further input of fossil fuel). Other conventional biofuel technologies typically deliver savings of greenhouses gases of less than 40 percent, compared to their fossil-fuel alternatives, which therefore may be insufficient atmospheric improvement to warrant extra strain on the terrestrial environment. 'When such impact as soil acidification, fertilizer use, biodiversity loss and toxicity of agricultural pesticides are taken into account, the overall environmental impacts of ethanol and biodiesel can very easily exceed those of petrol and mineral diesel.'[10]

The eventual EU legislation agreed in December 2008 took many of these considerations into account. For a biofuel to be counted towards a member state's 10 percent minimum

9   EEA Committee press statement, 10 April 2008, see also *www.eea. europa.eu*

10 OECD 2007 report cited above, p. 5.

renewable energy share in transport fuel, it must save at least 35 percent of greenhouse gas emissions compared to fossil fuels. The GHG saving threshold for target-qualifying biofuels will rise to 50 percent from 2017 onward, and from that date new installations must produce biofuel with emissions at least 60 percent lower than fossil fuels. The chosen fossil fuel benchmark for judging for GHG savings is the fairly tough one of Middle East oil, whose relatively easy extraction and refining requires little fossil fuel input (by contrast, virtually any biofuel would show enormous GHG savings if compared to, say, oil from Canadian tar sands).

The legislation sets out, for various biofuels, 'default' GHG savings rates which are generally below 'typical' rates. The default rate is the emission saving that a biofuel will be assumed to produce, in the absence of any evidence to the contrary. But, if they take the trouble to do so, producers can generally show to the EU authorities that their manufacturing technique will produce higher GHG savings, approaching the typical rate for that particular biofuel.

The table below has some examples of estimated GHG savings for different biofuels taken from annexes to the legislation. It illustrates, with the example of wheat ethanol, that the process fuel in making a biofuel can be crucial. It shows that corn or maize ethanol, the biofuel staple in the US, makes reasonable savings, but would only just meet the 50 percent EU threshold from 2017 on. It underlines that sugar crops produce a good GHG reduction, but that sugar cane (as grown in Brazil for instance) outperforms EU-grown sugar beet. It highlights that rape seed diesel, currently a European staple, may struggle under the new legislation to count towards national or EU targets. It points to the savings gained in using waste product, such as vegetable or animal oil, that has already been refined, or simply using gas as gas in the case of biogas from municipal organic waste being used as compressed gas to power vehicles. Finally, with the last three categories, it estimates the savings to be made from so-called second generation biofuels made from non-food crops.

But note that the chart below assumes that there has been no net increase in carbon emissions as a result of the change in

Table 13: Not all Biofuels are Equal

Biofuel production pathway (on the assumption of no net carbon emissions from land use change)	Typical greenhouse gas emission saving*	Default greenhouse gas emission saving*
Sugar beet ethanol	61 %	52%
Wheat ethanol (process fuel not specifed)	32 %	16%
Wheat ethanol (natural gas as process fuel in CHP plant	53 %	47 %
Wheat ethanol (straw as process fuel in CHP plant)	69%	67%
Corn (maize) ethanol EU-produced (natural gas as process fuel in CHP plant)	56 %	49 %
Sugar cane ethanol	71 %	71 %
Rape seed biodiesel	45 %	38 %
Sunflower biodiesel	58 %	51 %
Waste vegetable or animal oil biodiesel	88 %	83 %
Biogas from municipal waste as compressed gas	80 %	73 %
(Future) wheat straw ethanol	87 %	85 %
(Future) waste wood ethanol	80 %	74 %
(Future) farmed wood	76 %	70 %

*   Greenhouse gas saving compared to oil from the Middle East

Source: Annexes to resolution adopted by the European Parliament, 17/12/08

use of the land on which the biofuels are grown. For there are types of land that would release such large amounts of carbon on being converted to biofuel cultivation that biofuel 'saving' could never make up the carbon loss from the original land use change. Top in carbon storage are wetlands, followed by forests, because of the foliage in both. According to the United Nations' International Panel on Climate Change wetlands on average hold 686 tonnes of carbon per hectare, forests 275 tonnes per hectare and grasslands 181 tonnes per hectare, compared to only 82 tonnes per hectare of arable land.

Obviously, maintaining land so good at capturing and storing carbon is essential. So the Commission has proposed that no financial support or compliance credit should go to biofuels grown on land that, as of January 2008, was classed as wetland, mature forest, undisturbed forest, protected nature zones or highly bio-diverse grassland. Green groups criticized the Commission for setting the cut-off date so late that many of the slash-and-burn tropical clearance schemes of recent years will get into the EU biofuels scheme. The Commission said it had considered pushing the cut-off date back to 2003, the date of the previous EU directive on renewable energy, or even to 1992, the date of the UN Framework Convention on Climate Change. But it said it concluded that January 2008 was the appropriate cut-off because only then did its sustainability criteria become clear, implying that any earlier cut-off might unfairly penalize biofuel producers working on different assumptions about Brussels' eventual attitude. It is a pity that the EU did not think more about sustainability at the outset of its biofuel policy.

Many of the objections to the first 'booze' generation of biofuels would fall away if a second generation could be developed from 'lignocellulosic' biomass, from farm by-products such as straw, from wood products and from pulp and paper processes. Use of these inedible raw materials would avoid direct competition with the food industry, though there would still be some environmental concerns about what might be called 'factory forestry'. Indeed some first-generation biofuels only make sense as a bridge – and a short bridge at that – to the next generation. 'One reason that first generation biofuels continue to be promoted as serious solutions to the twin challenge of climate change and energy security is the notion that they will soon be supplanted by more advanced technologies now in development', according to the OECD study.[11]

But the same report goes on to cast doubt on whether second generation biofuels will become economically viable any time soon. It bases part of its doubt on logistics, not science. 'The logistical challenge of transporting biomass material to large production facilities is likely to impose a floor below which production costs cannot be lowered. This leads some to believe that the second generation biofuels will remain niche players, produced mainly in plants where the residue material is already

---

11 Ibid.

available in situ, such as bagasse (cellulosic residue from sugar cane pressing) and wood-process residues.' Such conditions are likely to be confined to Brazil and Finland.

Biofuels in moderation

For some years, however, the EU will have to make do with the current set of biofuels and cope with the dilemmas they cause. Having a mandatory biofuel target at some level is not a bad idea; the 2003 voluntary target produced little progress. Equally, putting too much stress on first generation biofuels is unwise, as many in the European Parliament pointed out. We have already seen in the previous chapter how Claude Turmes, the ponytailed Green MEP from Luxembourg, had a considerable influence on renewable electricity as the European Parliament's rapporteur on renewables legislation. On biofuels Mr Turmes wanted no mandatory target at all to encourage first generation biofuels. In the end he failed to get the 10 percent target killed, but he was very instrumental in scaling back incentives for first generation fuels by giving such favourable weighting to second generation biofuels.

One important advantage of scaling back the target, bringing demand more in line with sustainable supply, would be to reduce the incentives for producers to cheat on environmental standards. This is particularly important outside the EU, where the sustainability of biofuel production will inevitably be harder to police than in Europe.

Trade in biofuels will grow. Indeed it should grow. At present it only accounts for around 10 percent of global biofuel consumption. This is almost certainly too small, given that the wide differences in biofuel production costs around the world ought to make a higher proportion of commerce beneficial to all. But the EU, like the US, is generally keen to protect its biofuels sector from imports.

One instrument of protection is technical. The EU prescribes an iodine threshold below that generally in the soya bean oil grown by the big North and South American soya producers, while the tendency of palm oil, produced in quantity in south

East Asia, to go cloudy and waxy in cold weather inhibits to some extent its use in Europe.

But Europe's other means of protection are tariffs. These are relatively low (3.2–6.5 percent) on biodiesel. But because the EU is by far the biggest world producer of biodiesel, imports of it are equally low, except bizarrely imports from the US because of a US export subsidy (which Brussels has been contesting). EU duty on ethanol is much higher, 39 percent on denatured (rendered unfit for human consumption) ethanol and 63 percent on pure ethanol. Nonetheless Sweden in particular has become a very big importer of Brazilian ethanol, by importing it as a product for blending with petrol and thereby paying a much lower duty on it.

The EU needs to strike a balance on biofuel trade. It needs to persuade the domestic EU biofuel industry that Europe cannot hope to meet even scaled-down biofuel targets without a reasonable level of imports. At the same time, it needs to persuade foreign biofuel producers that they cannot hope to get into the EU market without observing environmental standards.

Neither task will be easy, as became evident at a biofuels conference that the European Commission hosted in Brussels in July 2007. While Swedish trade minister Sten Tolgfors argued that biofuel trade needed to be freed of all distortions so as to use 'the full potential of the international trading system to halt global warming' (and presumably to let Sweden import Brazilian ethanol duty-free), Ramon de Miguel, president of the European Bioethanol Fuel Association, claimed his industry continued to need import protection. Otherwise, he claimed, imports from countries like Brazil would jeopardize European investment in the sector, especially important research into second generation biofuels, and would undermine the extra energy security that home-grown fuels were beginning to offer Europe.

Most non-European biofuel producers at the Brussels conference grasped the need to convince their customers of the environmental acceptability of their product. President Luiz Inacio Lula da Silva said most of Brazil's sugar cane (for ethanol) was being grown far from the Amazon rain forest. While most biofuel producers stressed that they were using marginal or waste land, Yusof Basiron, chairman of the Malaysian Palm Oil Promotion

Council, acknowledged that some of his country's palm oil was grown on prime land. But he said that this had been opened to farming so long ago – as far back as 1917 – that it made no recent difference to the climate.

But Argentine farm minister Javier de Urquiza warned against the imposition 'from the outside' of sustainability standards. EU attempts to impose its standards unilaterally may be resisted by many countries, but mutually agreed international standards for certification of 'good biofuels' would also be difficult to negotiate. It would raise the tough issue – as with carbon compensation measures discussed in Chapter 10 – of whether it is possible and legal in international trade to discriminate, on environmental grounds, between *processes*, not just between *products*.

In sum, then, some further increase in biofuels is needed as the only way of tackling road transport emissions pending the commercial development of electric or hydrogen fuel cell cars. Some degree of compulsion is necessary to achieve this increase, because the biofuel share in road transport fuel is still only 1 percent despite sizeable tax exemptions in more than half EU states as well as quota obligations in some countries. Why should there be compulsion at the EU level? One reason is the alternative of increasing the biofuel tax exemption and spreading it across all 27 states would be hard to agree politically, and create a very uneven instrument, given the lack of any common EU level of fuel tax that biofuel would be exempt from. The broader reason for common action on biofuels is to avoid distortion in Europe's internal and external markets. Some imports are vital to prevent environmental damage in Europe. But some sustainability standards are vital to prevent environmental damage outside Europe.

In retrospect, the EU should have established its environmental criteria for biofuels some years ago in less politically charged circumstances. If the debate becomes too polarized, it condemns the EU to the kind of inaction, which, as we shall see in the next chapter, is evident with nuclear power.

CHAPTER 13

NUCLEAR POWER: THE IMPOSSIBLE CONSENSUS

The EU needs to spend at least 30 times more on nuclear waste management research.

Loyola de Palacio, European energy commissioner, 2003

Earlier, in Chapter 2, nuclear power was rated as having a high potential for EU collective action. This does not necessarily mean a common policy, which would be impossible when 13 member states do not have, and some of those do not want, nuclear power. It is rather the EU's potential ability to make nuclear power development easier for member states than it would be if they did not belong to the Union. As mentioned in Chapter 2, nuclear power's EU potential is rated at least as high as that of energy market policy, because it was given a complete institutional framework right from the start with the Euratom treaty of 1957.

But, in spite of Euratom, all key nuclear power decisions are national and are likely to stay so for a long time. Nor has a lavish EU nuclear research programme solved the problems that most worry Europeans about atomic power, such as final disposal of radioactive waste. So it is easier to argue that nuclear power has contributed more to Europe than Europe has to nuclear power.

The past year, 2007–8, has seen a modest revival of EU-level interest in nuclear power. The Commission has formed new groups of national regulators, officials, executives and researchers to discuss how to improve safety and radioactive waste management, how to harmonize national rules in these areas with a view to reducing the differing national standards that new reactors would have to meet across Europe, and how to make regulation and risk in nuclear power more comprehensible and hopefully acceptable to mostly sceptical European publics.

Such efforts are appropriate. For nuclear reactors still generate a third of total electricity in the EU. This contributes to
Europe's energy security; though natural uranium is almost
entirely imported, it is a small part of nuclear power's total
cost and its enrichment into reactor fuel is mostly carried out
within the EU. It contributes even more to the fight against
climate change. 'Continued use of nuclear energy in the EU is
almost certainly going to be necessary to attain the policy goals
in climate change and security of supply', said the International
Energy Agency in its 2008 report on EU energy policy. The
EU's 152 reactors, more than in any other region of the world,
provide two thirds of Europe's carbon dioxide-free power As
such they draw indirect financial benefit from not needing the
carbon emission permits now required of electricity generators
using fossil fuels. So the perspective for nuclear power in Europe
should be brighter.

Yet nuclear power will be lucky to maintain its one-third share
of generation in the future. The average age of the EU's 152
reactors is around 25 years. This would not matter – reactors
are typically designed for a 40-year working life that these days
can be safely extended by a few years – if a reasonable rate of
replacements was being planned.

At the time of writing, only two reactors, one in Finland and
one in France, were being built inside the EU. More countries
are considering building new reactors. They include the three
new member states: Lithuania, Slovakia, and Bulgaria, which are
being obliged to shut down, as a condition of their entry into the
EU, Soviet-era reactors judged to be unsafe. But there is interest elsewhere in Eastern and Central Europe, a region where
green political parties are weak and where economic factors still
tend to prevail over environmental ones, in expanding nuclear
power. Romania and the Czech Republic plan to expand their
atomic power programmes, while Estonia, Latvia and Poland are
discussing participation in a new Lithuanian reactor.

In Britain, the first country in Europe to open a power reactor, the Labour government has decided in favour of replacing
its existing reactor fleet (the oldest in Europe with an average
age of 30 years), but is leaving to the market the question of
how and who should do this. Again, at the time of writing,

the UK government had yet to entice any company into a firm contract to build new reactors. In Italy, the first country in Europe to abandon its nuclear programme (after a 1987 referendum with a narrow No majority vote against nuclear power), the government of Silvio Berlusconi, re-elected in 2008, announced it would seek to reintroduce nuclear generation. This will not prove simple.

Having shut down its programme, any government in Italy may feel the lack of local support for nuclear power surrounding reactors that are still operating, though there is still some residual employment around old reactors to carry out decommissioning. For it is an observable fact, virtually everywhere, that the strongest backing for nuclear power comes from those most immediately living and working with it; this is why the easiest place to put a new reactor is next to an old one. Any such difficulty in restarting nuclear power from scratch may give pause to those other EU states that have said they will phase out nuclear power by not replacing their existing reactors. Germany, Spain and Sweden are still on course to do this eventually, although there is a debate in each of these countries about the wisdom of this. Belgium announced in spring 2008 that it was reviewing its gradual rundown of nuclear power.

So the situation differs country by country, and it hard to see how things could be otherwise, given present perceptions of the costs and benefit of nuclear power. The memory of the 1986 Chernobyl reactor accident in neighbouring Ukraine is still too recent, and the prospect of catastrophic climate change, thankfully, still too remote, for EU states to agree – as they have on renewables – that nuclear power should form a *common* part of their energy mix.

Moreover, different European societies cope differently with the challenges of nuclear power. For France, the military use of the atom is not seen as original sin tainting civil nuclear; indeed its nuclear arsenal, the *force de frappe*, is a source of national pride. France, too, took very seriously the energy insecurity it felt during the 1973–4 oil shock, and as a result now has 59 of those 152 EU reactors. Public attitudes in other European countries to nuclear power have been shaped by issues such as reactor safety and particularly the lack of long-term nuclear waste disposal. Not surprisingly, people are reluctant to contemplate new reactors producing new waste before ways have been found to deal with old waste from old reactors.

After long consultation and a very Nordic process of consensus-building, Finland took the decision on a long-term burial site for nuclear waste and, then, felt able to decide to build a new reactor nearby. This contrasts with France, which has still not decided on a final waste depositary but whose society is evidently ready to follow the state's lead in nuclear matters. The UK is somewhere in between France and Finland. As Malcolm Grimston has put it, Britain is 'more market-oriented but with governments which seem confused as to whether markets and consultation (on the one hand) or central diktat (on the other) are the appropriate mechanism for managing the interface between science and society'.[1]

What is clear is that the EU has done little to make individual countries' decisions about nuclear power any easier, let alone fashion a common policy. This is despite having, from the outset in 1957, the Euratom treaty which aims to promote nuclear energy in general and in particular uranium fuel supply, operational safety in reactors, safeguards against weapons proliferation, nuclear research, and has, in addition, a large staff to carry out these tasks.

Proliferation safeguards

Since 1970, when the Non-Proliferation Treaty (NPT) entered into force, the task of policing this United Nations treaty has been left to the UN's anti-proliferation body, the International Atomic Energy Agency (IAEA). But by that time, Euratom had already developed its own elaborate safeguards to prevent the spread of nuclear weapons, not least because in 1957, with Germany only recently allowed to rearm and join NATO, France wanted Euratom designed to keep an eye on any German atomic activity. (French desire to monitor Germany also lay behind the

---

1   Malcolm Grimston, 'The importance of politics to nuclear new build', Chatham House report, 2005, p. 42.

Euratom requirement that the Commission must be informed of any nuclear investment, and periodically the Commission publishes the state of EU nuclear investment plans to give them more transparency.)

Since bureaucracies are loath to renounce any rationale that keeps them in being, Euratom inspectors have kept on inspecting EU reactors, just as IAEA inspectors do. Duplication in so sensitive an area may not be an expensive luxury in this era of terrorism. And it is true that the Commission has recently scaled down the Euratom inspectorate, which numbered 180 people in 2006, as the result of more coordination with the IAEA. But even in the one area of inspection where there is no overlap with the IAEA, one has to wonder at Euratom's purpose.

In contrast to the IAEA, which has to respect the special privileges that the UN Security Council's five weapon states have under UN law in the NPT, Euratom has the right, under EU law, to inspect the civil installations of Europe's two weapon states, Britain and France. Euratom prides itself on this. Yet what precisely is there to safeguard here – apart from preventing outside theft of nuclear material which one must assume to be also a British and French concern – when Britain and France have openly (and in terms of UN conventions, legally) turned atomic material into bombs?

Fuel supply

Euratom's supply agency was created to ensure member countries' reactors got fair and regular access to nuclear fuel. The original expectation, at a time when there were relatively few sources of natural uranium and when the US was virtually the only provider of enriched uranium, was that it would need to manage a shortage of supply. In reality, there has been abundance, and as a result, the supply agency's role has moved from one of promoting imports to controlling them.

Nuclear supply contracts within the EU need Euratom approval. Indeed, in its early years Euratom used to be the co-signatory on all nuclear supply deals (except for France's contracts with Niger and Gabon, which Paris regarded as its private preserve). The big change in the market, for Europe, came in the early 1990s when, after the Cold War, Russia started to offer large quantities of natural and enriched uranium to the European market. Natural uranium imports pose no competitive threat to the EU where natural uranium is not mined. But enriched uranium imports do compete with enrichment plants in the EU. In particular, low cost Russian enriched uranium was judged to threaten relatively high cost EU enriched fuel, especially that made by Eurodif in France and by Urenco in the Netherlands.

As a result, when the EU signed its Partnership and Co-operation Agreement (see Chapter 9) with Russia in 1994 on the Greek island of Corfu, it adopted on the side a unilateral declaration on imports. The Declaration of Corfu, which has never been formally published, is to the effect that the market share of EU uranium enrichers should be maintained at around 80 percent, for reasons of security of supply. (The principle of setting a limit on imports was also confirmed for natural uranium, but for the reason mentioned above, this was less sensitive).

This effective limit on imports from Russia to 20 percent of the EU market has since bedevilled EU-Russia relations. As the Commission noted in 2002, 'every official meeting, including EU–Russia summit meetings, is treated as another opportunity for the Russians to protest about [nuclear fuel] restrictions and to call for a satisfactory resolution on trade in nuclear materials.'[2] In the last few years the restriction has come to bother Moscow slightly less because it has been making so much money selling oil and gas to Europe. But the issue has not disappeared as an irritant in EU-Russian relations, and will certainly re-emerge in any negotiations for a new EU-Russia agreement.

However, it is not proliferation fears or fuel worries that deter EU countries from developing or maintaining nuclear power, but rather the issues of operational safety, waste disposal and reactor decommissioning that worry their voters.

---

2   Commission Communication on nuclear safety in the EU, COM (2002) 605 Final, pp. 5–6.

Safety

Euratom sets basic standards for radiation protection for people working in reactors. But bizarrely it has no role in setting safety standards for the design or operation of reactors, when for the population at large the risk of radiation exposure comes from faulty reactor design or operation. EU member states have always insisted on regulating their own nuclear installations, and setting their own reactor safety standards. 'Nuclear safety and radiation protection are now two closely linked concepts serving a common health protection objective', complained the Commission in 2002. 'Consequently it is now no longer possible or desirable to separate these two disciplines.'[3] But nothing has changed since 2002.

EU countries have, often to their detriment, had their own special ideas about reactor design. The UK is the classic example, where special UK-only designs for the Magnox and Advanced Gas-cooled Reactors (AGRs) have made impossible the sharing of economies and lessons with other countries and other nuclear programmes. A current instance of national particularism in reactor design is the European Pressurized Water Reactor (EPR) which the French-led consortium is building in Finland, and whose cost over-run and delay is partly due to design changes demanded by the Finnish regulators. This is unfortunate in the sense that TVO, the Finnish power company ordering the reactor, has been collaborating with other European companies to develop an industry-led harmonization of reactor design in the 'European Utility Requirement' (EUR) initiative. And TVO had used EUR as the bid specifications template for its new reactor.

It is not surprising that nuclear regulators differ, for they often have not only different ideas of what is safe, but different ways of arriving at those ideas. 'The Germans traditionally took a very prescriptive approach to reactor design,' notes one EU expert, 'whereas the UK and French regulators have tended to leave it to the companies to prove a design is safe.'[4] Lack of a common

---

3   Ibid, p. 7.
4   Author interview 2008.

standard or design across Europe obviously poses problem for any reactor manufacturer trying to gain economies of scale in replicating the same model.

In recent years, what evolution there has been towards common safety standards has come from the work – outside of the EU and Euratom – of the Western European Nuclear Regulators Association (WENRA). But in 2007, the Brussels Commission took a more proactive approach by setting up the European High Level Group on Nuclear Safety and Waste Management. Mainly composed of the EU's 27 national nuclear regulators, this body may in time produce a common approach in its two areas of responsibility.

Waste disposal

This is the issue that most exercises people about nuclear power. Indeed, according to a July 2008 Eurobarometer opinion survey, 40 percent of opponents of nuclear energy said they would *change* their mind if some safe and permanent solution could be found for radioactive waste. In the same survey, more than 60 percent of respondents wanted an EU role in monitoring national management plans for radioactive waste, and felt such national plans should be required and harmonized across Europe.[5]

Governments are often asked, 'How can you possibly decide to build new reactors when you have not decided what to do with waste from existing ones?' So far, only one has come up with an answer. Finland only decided to go ahead with building its latest reactor after it had decided on a final geological depository for nuclear waste, the first country to do so.

On so sensitive a matter, the EU or Euratom would be ill-advised to tell countries what to do with their nuclear waste, provided its basic conditions for radiation protection are met. Nor can it dictate the timetable for countries to decide on waste disposal. But what the EU could have done is put more effort into researching ways of permanently and safely storing nuclear

---

5   Eurobarometer surveys, *http://ec.europa.eu/public_opinion*

waste that governments could draw on, particularly if they are contemplating building new reactors.

One measurement of effort is money. In Euratom's 2002–2006 sixth framework programme, a mere Euros 90m was devoted to radioactive waste management, plus part of the Euros 290m nuclear research budget of the EU's Joint Research Centre. Compare this, however, with the Euros 750m that went over the same period to research into fusion. This distant dream of fusing atoms to reproduce the energy of the sun continues to eat up EU research money. In the seventh framework programme for 2007–13, of the Euros 2.75bn going on nuclear research, fusion will get Euros 1.947bn, compared to Euros 287m on solving the rather more immediate problems and issues of fission and radiation protection.

'I think the money for fusion should be more calibrated, because fusion is always 40 years away in the future and focusing on fission is more realistic', says Santiago San Antonio, director general of the Foratom nuclear industry association.[6] He points to the creation in 2007 of the Strategic Nuclear Energy Technology Platform to research the fourth generation of fission reactors as the sort of 'recalibration' he wants to see.

It should, however, be said that some EU nuclear experts believe that the EU should not necessarily be spending more money on more basic research into waste disposal, but rather on applying the techniques that are known to the potential burial sites. 'We are not really now doing basic research on waste, and we don't need to', says one expert. 'What we do need is more *in situ* testing, to get digging, looking at the geology of potential sites and doing things like heat tests on the rocks.'[7] However, such *in situ* work generally requires sites to be chosen beforehand by EU countries. This creates a circular chicken and egg problem, with the EU only able to provide the research that would help states choose storage sites if the states have already chosen the sites.

---

6   Author interview, 2008.
7   Author interview, 2008

An attempt at change

Euratom – or the European Commission into which Euratom's secretariat was subsumed in 1968 – did have a serious try at reform in 2003. In January of that year, the EU executive proposed draft directives on common rules on reactor safety and for funding the decommissioning of reactors, as well as on an obligation on member states to set a timetable to bury their radioactive waste.

This was not exactly a bolt out of the blue. The directives were partly drafted to reflect in EU legislation two international conventions that had already been adopted by most member states, after a typical EU institutional fight over treaty competences. In 1996 EU states negotiated within the IAEA the Nuclear Safety Convention. This did not create any requirement for a European safety standard, because there was the IAEA one. But when the Commission proposed that the EU, as Euratom, become a party to the convention, a number of member states objected, and the Commission took the issue to the European Court of Justice. While the appeal was being considered, EU states negotiated another convention in the IAEA, the Joint Convention on the Safety of Spent Fuel Management and on the Safety of Radioactive Waste Management, which the Commission also wanted Euratom to sign. Eventually, the European Court of Justice ruled in 2002 that the EU had competence in nuclear safety, and so Euratom got to sign the conventions.

But the real catalyst for Commission action at the turn of the century was, in a way, the very event, the 1986 Chernobyl accident, that had cowed the Commission into silence on nuclear matters through most of the 1990s. As the prospect neared of enlargement and of East European states bringing their Chernobyl-style reactors into the EU with them, so concern about nuclear safety grew. In the late 1990s the Commission started to negotiate the closure of the riskier reactors from candidate countries, but found itself bitterly criticised by East European governments for having no proper criteria – because no EU safety standards – by which to judge them.

Eventually, the late Loyola de Palacio, a feisty Spanish conservative who was commissioner for energy as well as transport, decided to exploit impending enlargement to East Europe to advance a more proactive EU nuclear power. So, armed also with the ECJ ruling that backed Euratom's competence in nuclear safety, she unveiled her draft directives in January 2003.

The proposals got the backing of the European Parliament, but were attacked from many other quarters. A majority of member states supported the Commission. This was not surprising, because nuclear safety issues were already figuring in EU summit communiqués.[8] But some governments regarded Ms de Palacio's proposals as a Commission power grab (which in part it was), and saw no problem in perpetuating double standards on safety, one for existing club members, another for newcomers. They also disliked, on subsidiarity grounds, Brussels involving itself in the details of decommissioning and waste disposal. There were enough objecting governments to form a blocking minority. Within this blocking minority, the UK was the most active. According to one official, 'the UK's strong objections were related to its worries that its Magnox reactors would not stand up to European scrutiny of safety, and to the fact that no decision had been taken in the UK, at that time, about geological disposal being the best way to go with radioactive waste material.'[9]

The nuclear industry itself was fairly supportive. It believed the directives would have pushed member states towards the harmonization of safety standards and towards decisions on waste management programmes that it, the industry, wanted in order to develop the sector. For their part, environmental groups believed, with some cause, that measures ostensibly designed just to increase nuclear safety had the wider purpose of revitalising an industry they oppose. For this reason they found themselves in the uneasy position of opposing clearer regulation on safety and on the necessity to publish nuclear waste management plans.

In September 2004, the Commission watered its proposals down, but not sufficiently to give them any chance of passage

8   The Laeken summit of 2001, for instance, said 'the European Council undertakes to maintain a high standard of safety in the Union'.
9   Author interview, 2008.

through the Council of Ministers. "Eventually we will need legislation on nuclear safety and waste", Dominique Ristori, the Commission's top nuclear official, said in mid-2008. "2003 was a bit premature [for agreement on the directives], but at the right moment we will come back with legislation which will be based on the fundamental rules already agreed internationally in the two conventions."[10]

In fact, the Commission judged 'the right moment' to revive one of its proposals was as early as November 2008, when it proposed a recast directive on nuclear safety.[11] However cynics might say the Commission's timing was mainly to satisfy France's desire for a nuclear proposal during its autumn 2008 spell in the rotating EU presidency. For the 2008 proposal was still weaker than the 2004 one, which had itself been diluted with the abandonment of an EU fund for reactor decommissioning. In 2008 the Commission stressed its recast proposal was to strengthen the role and independence of national nuclear regulators, who had played a big part in opposing its earlier draft directives. Virtually the only *communautaire* element in the 2008 proposal was the requirement that national nuclear regulators submit every 10 years themselves and their national systems to 'international' (left unspecified) peer review. Meanwhile, the 2003 directive on nuclear waste disposal, proposed in 2002 and revised in 2004, remains in limbo.

Nonetheless, the Commission has at least tried to dissuade more governments from abandoning nuclear power. In its spring 2006 green paper, the precursor of the present strategy, it reminded member states that while they were free to choose their energy mix, 'decisions by member states relating to nuclear energy can also have very significant consequences on other member states in terms of the EU's dependence on imported fossil fuels and CO_2 emissions.' The green paper made exactly the same point about the need for member states to consider the effect on the wider Union of their decisions on gas. 'Decisions to rely largely or wholly on natural gas for power generation in any given member states have significant effects on the security

10 Author interview, 2008
11 COM (2008) 790/3.

of supply of its neighbours in the event of a gas shortage', it warned.[12]

Italy is a case in point on both counts. Having closed down all its nuclear power plant at home, it imports very large amounts of gas and runs a chronic deficit in electricity trade, including imports of French nuclear-generated power. These are evidently the sort of considerations that the Commission wishes the Italian people had taken into account in 1987, when, in the immediate wake of Chernobyl, they voted by referendum to shut the country's nuclear power plants.

De Palacio's successor as energy commissioner, Andris Piebalgs, has been more circumspect in promoting the EU dimension in nuclear policy. On a purely personal level, this would be understandable. A Latvian who was a Soviet citizen in 1986, Mr Piebalgs was kayaking in Ukraine at the time of the Chernobyl and only found out about it two days after it occurred. Nonetheless, as energy commissioner, he has accepted nuclear power's essential role in climate change policy. In 2007, he has oversaw the creation of the High Level Group of national regulators to discuss safety and waste management, and the opening of the European Nuclear Energy Forum as a talking shop that will regularly alternate between Bratislava and Prague. The willingness of both the Czech and Slovak governments to host this forum is a sign of East European countries' seriousness about nuclear power. They tend to regard it as a surer road to a low-carbon economy than renewable energy.

Certainly the economics of nuclear power are better than for some time. Uranium prices rose 10 times in the 2003–7 period, though they have fallen back since. However, even at its price peak uranium still counted for much less in the cost of electricity it generates (because of the very high capital cost of reactors) than fossil fuels do. Nuclear power cannot for the moment expect the same overt public subsidy that goes to renewables. Nor does Mr San Antonio of Foratom believe it needs any state aid, 'just a stable framework over a long period in which to recover the investment'. Thinking of his own country, Spain,

---

12 Green Paper, 'A European Strategy for Sustainable, Competitive and Secure Energy', 2005, p. 9.

he says, 'nuclear energy cannot be a political football at every four-year election.'[13]

But nuclear power operators now reap advantage from the system of carbon permits (traded on the Emission Trading Scheme) that penalizes rival generators using fossil fuels. So, for the first time in many years, the EU dimension is making a real contribution to nuclear power. The technology-neutral characteristic of the ETS, which rewards all low-carbon technologies alike, is one of the beauties of the system.

It is not surprising that nuclear power's main financial assistance these days should have to come in rather disguised form through the ETS. For over the past 50 years, the consensus that once existed in favour of nuclear power has evaporated. How can 'Europe' actively promote nuclear power, when people throughout the EU institutions – the Commission as well as the Parliament and Council of Ministers – are split over it? Because Euratom is a founding treaty of the EU and was signed long before opt-outs were created to cater for awkward members such as the UK, it has been part of the set menu for all, something that all new members have signed. This may have been a mistake. The EU now has some viscerally anti-nuclear members, especially Austria and to some extent Ireland, which are full members of Euratom. They are against nuclear power not only for themselves but also for others. They can be an obstruction to progress, just as Britain would have been had it been forced into the common eurozone currency zone or the Schengen free-travel area.

An example of Austrian obstructionism came in February 2008 when Austria threatened to block energy ministers' agreement on the new Strategic Energy Technology plan unless it carried a guarantee that no money would go to nuclear research. A temporary compromise to please the Austrians was found whereby the SET plan was approved, but the financial consequences of that approval were left to be decided later.

In these circumstances, one can argue that Austria should abstain, or opt out, rather than obstruct research that might one day make nuclear power palatable even to Vienna itself. Indeed,

---

13 Author interview, 2008

why not have in the nuclear field the sort of variable geometry successfully tried in other areas, such as the euro currency zone, Schengen or EU defence? This would allow the Austrias and Irelands to opt out of Euratom, and turn Euratom into a sort of 'coalition of the nuclear willing'.

The snag is that no one wants to opt out of Euratom. As long as it exists, Austria and Ireland want to participate, if only to keep an eye on their neighbours' nuclear power plans. There was a moment in 2004 when change seemed briefly possible. During negotiations on a new constitution, five countries – Germany, Ireland, Hungary, Austria and Sweden – declared their interest in an intergovernmental conference to review the terms of Euratom. But they found no wider support, and the issue was dropped and is likely to stay dropped. Subsequent events with the Treaty of Lisbon have shown that EU treaty negotiation, and especially ratification, is contentious enough without adding in the nuclear power issue.

In reality, countries that are undecided about nuclear power may be more of an obstacle to Europe's low-carbon energy development than the outright opponents to it. In the undecided camp must be counted those countries – Belgium, Spain, Germany and Sweden – which have agreed to phase out nuclear power, but necessarily over long periods that provide opportunities for politicians to change their minds. Such countries fall between two stools. They make no plans to build new reactors, but as long as the possibility of a U-turn exists, they also shy away from committing themselves absolutely to replacing all their nuclear power with alternative energy. Germany and Spain have increased their renewable energy enormously, but not by enough to fill the energy vacuum that phase-out of their reactors will leave. How to plug this nuclear vacuum in the future is one of the challenges for Europe's energy research and development programmes to which we now turn.

CHAPTER 14

ENERGY R(ELUCTANCE) AND D(ELAY)

This market gap between supply and demand is often referred to as the 'valley of death' for low carbon energy technologies.

The European Commission on launching its energy technology plan in November 2007

Develop a new technology in every other sector of the economy, and you will usually have a market for it. As long as it does something new – not even necessarily useful (think of kids' electronic games) – or does something old but in a cheaper or better way, then you will have ready customers. Not so in energy. There seems to be an inbuilt lack of market interest in new energy technology that makes energy innovation especially difficult.

The problem is not a question of long lead times (except in nuclear fusion which always seems to be 40 years from commercialization). It is only partly the network challenge of connecting new energy sources to grids or transforming grids to suit decentralized power sources so that energy reaches everyone. Mainly, the problem is that low carbon electricity technologies are almost always more expensive than those they replace, but provide nothing more than the same old electrons. Equally, carbon capture and storage (CCS), which is another way of keeping carbon out of the atmosphere, is in a sense an assault on productivity, a step backwards. CCS is perfectly justifiable, because it is for the greater good of the planet, but is nonetheless a technology that has the effect of reducing the electricity output of the average power plant to what it was some years ago. This is because the process of capturing the CO_2 and pumping into underground storage itself requires power.

Energy efficiency measures, examined in Chapter 15, have a payback in lower energy bills. But in the case of households, this may accrue to the advantage of tenants rather than landlords who took the measures in the first place. In general, the benefits of most low carbon technologies often flow more to society than to the buyer.

Moreover, the innovation process, the introduction of low carbon energies has to take place in energy systems that have been optimised over decades. Yes, the lights went out across Italy in 2003 because a tree fell on a pylon in Switzerland, and across a wider area of north Central Europe in 2006, when a Germany utility had to switch off a power line to let a ship on the Rhine pass underneath, and other power lines became overloaded. Yes, too, for a few hours at the beginning of 2006 gas flows were reduced to several European countries as a result of a Russian dispute with Ukraine. But energy supply, at least in Western Europe, is reliable for 99.99 percent of the time.

Climate change, however, has altered the old order. In the past there has certainly been a close correlation between the oil price and energy R&D (see the chart below). When the oil price came down in the mid-1980s, so did energy R&D and it has only recently picked up. Had the 1980 peak in energy R&D been sustained, the situation would have been different because the EU and its member states would collectively now be spending Euros 7–8bn a year on energy, instead of Euros 2.5bn. But only a few countries took the 1970s oil scares seriously – these exceptions were France with its nuclear programme, and Denmark and Japan with their big investment in energy efficiency. Fifteen years of fairly low oil prices, from 1986 to the turn of this century, left the EU as a whole with, in the words of the Commission, 'accumulated under-investment [in energy R&D] due to cheap oil'.[1]

That period is over. Because of carbon constraints, we can no longer rely on oil prices triggering sufficient levels of research into alternative energy. Because of carbon constraints, we need to hurry on with this research regardless of what the oil price does. In any case, the oil price can no longer be trusted as a prop for energy R&D spending. It bounces around too much to be a reliable prop.

1　Memo/07/469, the Commission, p. 1.

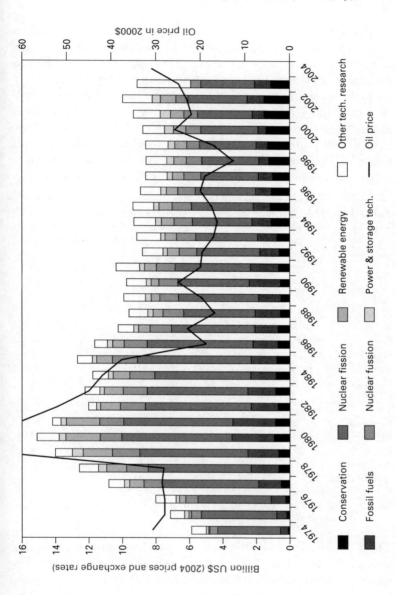

Figure 7: Energy R&D Spending in OECD Countries and the Oil Price

Source: European Commission Memo/07/469

The corporate energy sector is no better when it comes to R&D, though some of the research cuts in recent years may stem from privatization of many European energy companies and the liberalization of some European energy markets. The table below shows how little some of the major European utilities spend on R&D (new technologies), as distinct from capital expenditure (expansion and maintenance). The table also shows that oil companies are equally frugal on R&D – although their capex is high – because they can rely on research-intensive service companies such as Schlumberger and Halliburton.

Table 14:   The Corporate Sector's Poor R&D Performance

Company	£Million	Research and Development Spending 2007		Percent of operating profit	Percent of sales
		Percent growth over			
		last year	average of last four years		
EdF	262	-3	-5	4.6	0.7
Endesa	26.98	-7	44	0.8	0.2
RWE	90.96	6	-52	2.3	0.3
Suez	57.94	1	-8	1.9	0.2
Gaz de France	56.60	15	-9	2.3	0.3
Scottish & Southern	6.30	350	530	0.6	0.1
Eon (UK)	5.0	67	100	0.5	0.1
RD Shell	452.18	51	61	2.0	0.3
Total	383.37	-16	-14	2.1	0.4
ExxonMobil	374.52	3	12	1.1	0.2
BP	201.82	-21	-5	1.3	0.1
Eni	149.58	10	-35	1.1	0.3
Schlumberger	316.43	23	14	11.9	3.2
Halliburton	141.53	15	14	8.1	1.2

Source: drawn from R&D scoreboard, UK Department of Business, Enterprise and Regulatory Reform

The EU has tried to play a part in remedying this situation. For reasons examined in Chapter 13, its energy research programme has been too tilted towards nuclear, and within nuclear too tilted toward fusion. But there is a wider energy research effort. The annual average devoted to energy in the EU's current 2007–13 research framework programme is Euros 886m, up from average Euros 574m a year in the EU programme for 2002–6. There is widespread acknowledgement, however, that too much of this has been increasingly scattered around in penny packets, because the trend has been towards smaller projects with more partners.

So, as part of its attempt to create a brave new world in European energy, the Commission came up in November 2007 with a 'Strategic Energy Technology' plan. It was specifically directed at low carbon technology 'for which there is neither a natural market appetite nor a short-term business benefit'.[2] In any other context or sector, the idea of Brussels backing technology it knew the market did not want would be anathema. But as we have seen energy innovation is especially problematic. The aim, in the words of one Brussels official, was to 'shepherd early energy technology through the so-called Valley of Death, which lies between the demonstration stage and getting big enough markets to survive'.[3] Further downstream, the Commission has a programme called Intelligent Energy. In the words of its director Patrick Lambert, it seeks to 'create market conditions for acceptance of new energy technology, such as designing EU-wide qualifications and courses for the training of installers of wind turbines'.[4]

The Commission has said the main EU technology goals over the coming decade are to:

- Make second generation biofuels competitive.
- Commercialize carbon capture and storage.
- Double the generation capacity of the largest wind turbines, especially for offshore use.
- Demonstrate the commercial readiness of large-scale solar power.
- Make possible a smart grid for Europe, able to take renewable and decentralized sources.
- Bring to market efficient energy conversion devices such as fuel cells for use in buildings, transport and industry.
- Improve the prospects of nuclear fission by solving the waste problem.

---

2 A European Strategic Energy Technology Plan, Commission communication, COM (2007)723 final, p. 3.
3 Author interview, 2007.
4 Author interview, 2007

These challenges are daunting enough to require a pan-EU effort. One of them – the smart grid – is also of a geographic scale that requires an EU effort. Indeed it is almost the technology equivalent of what the Commission is trying to achieve through liberalization and market integration.

In energy research, the EU, through the Commission, generally has two useful roles. One is convening and coordinating. Not every EU state belongs to the International Energy Agency, which generally requires that its members must first join its mother institution, the Organization for Economic Cooperation and Development. Only those EU states that do belong to the IEA (19 out of 27) have a dedicated forum in which to discuss energy research. Now, there is supposed to be a EU steering group, chaired by the Commission, to coordinate EU and national research efforts; a series of European Industrial Initiatives in the form of public-private partnerships in specific technology areas; and a European Energy Research Alliance linking universities and focusing more on basic energy science. The other use of the EU is its role in setting technical standards for its huge single market. 'Standards are a competitive element these days', says an outside specialist. 'No European company can afford not to bring a European standard to the table when it is, for instance, talking to the Chinese.'

But such powers of convening, coordinating and standard-setting are not enough to galvanize Europe into giving the world a lead in demonstrating the technical and economic feasibility of carbon capture and storage (CCS). This technology will take time to prove commercially, but is considered a vital contribution to preventing emissions from spiralling up in the 2020–30 decade before advanced renewable energy and revived nuclear power can take them down to much lower levels. The EU is keen to lead, by example, China and India, with their huge coal reserves, into adopting CCS technology. Yet in the January 2008 climate change package, the mismatch between the Commission's ambition and means was especially glaring in CCS.

Likely costs and benefits of CCS are both big. By capturing carbon dioxide as it comes out of power stations, funnelling it underground (most likely depleted oil and gas fields) and keeping it there, CCS technology could reduce emissions in the EU by 13 percent of total power and steam generation emissions by 2030.[5] Failure to act soon might have larger negative consequences. One Commission study estimated that if the EU were to delay CCS demonstration technology for seven years, and if this led to the same delay around the world, this could mean over 90 Gt [gigatonnes] of avoidable CO_2 emissions being released worldwide by 2050. This would amount to 20 years of total current EU emissions.

Costs are high too. The bill for research into CCS might not be that large – in all Euros 1bn between now and 2020 – but the industrial costs would be on the same scale as nuclear fusion, running into billions. To prove various CCS technologies in various geologies in various places around Europe, the Commission has proposed, and EU leaders have agreed, that a dozen demonstration plants need to be up and running by 2015. The present experience in and around Europe with CCS is limited to Statoil's extraction of CO_2 from its Sleipner field (due to Norway's high CO_2 tax) and to a BP-Sonatrach project in Algeria (motivated by BP's internal carbon trading scheme).

Installing the capture, transport and storage equipment would add anywhere between 30 percent and 70 percent in up-front investment to the cost of a standard power plant. Moreover, operating costs of CCS plant would probably be 25–75 percent more expensive – mainly because of the power diverted to running the CCS equipment – than with non-CCS coal-fired plants. Climate Change Capital, the specialty investment bank, calculated in 2007 that the dozen CCS demonstration plants would need financial support of Euros 1.5bn–4bn a year or Euros 10.3bn–16.4bn in upfront grants.

This scale of money is out of the EU research budget's financial league. At one point in 2007, the Commission's energy division had hoped to divert some serious EU money, coming from unspent farm funds, into CCS development. In the end this money went to rescue the Galileo navigation satellite project. So when the Commission set out in January 2008 its draft directive for a regulatory framework for CCS deployment, it had no more

---

5   Supporting Early Demonstration of Sustainable Power Generation from Fossil Fuels, Impact Assessment, SEC (2008) 47, p. 35.

financial aid to offer than a proposal that safely stored CO_2 should be treated under the emissions trading scheme (ETS) as not emitted. This would mean that a CCS operator would not have to buy ETS allowances as his non-CCS rivals would have to. But the Commission admitted this incentive would be insufficient until the cost of avoiding carbon through CCS was equal or lower than the cost of emitting it with an ETS permit, and it acknowledged that this crossover point was unlikely to occur before 2020.

In terms of upfront investment money in CCS, the Commission said it was counting on government state aid and corporate finance. It said it would take a very benevolent view of state aid to CCS, but there has been little so far. Only a few governments have been come up with any firm aid promises. Among them are the UK, which has invited companies to compete for a grant to develop a relatively small CCS plant (300MW), and Norway. The latter, though not in the EU, belongs to the European Economic Area and has to abide by EU internal market rules, including state aid. In July 2008, Brussels happily allowed Oslo to put some of its oil riches into funding up to 80 percent of Norway's Mongstad CCS project.

To the companies, the Commission held out a weak mix of carrots and sticks. It appealed to companies' self-interest in gaining a first-mover advantage in CCS, and offering itself to give 'first movers a means of coordination, exchange of information and identification of best practices'. It ventured, rather endearingly, the notion that giving 'a European logo' to CCS projects might be an extra inducement for industrialists to part with several hundred million euros. But 'without bold funding decisions by the companies at the earliest opportunity, complementary public funding may not be triggered', it warned. It then managed to make a threat, and then in the same breath, withdraw it. 'The longer the power industry takes to start embracing the CCS technology, the more policy-makers will be obliged to look at the option of compulsory application of CCS technology as the only way forward.' But the Commission's own impact assessment acknowledged that the risk of imposing commercially unproven technology on the sector could not be justified.

Industry has also played a game of financial bluff and bluster. An early European Technology Platform was created to develop CCS under the name of Zero Emission Fossil Fuel Power Plants (ZEP). Even after it was clear that the Commission's financial cupboard was bare, a group of some 25 utilities, oil and engineering companies belonging to this ZEP programme wrote to Mr Piebalgs, the energy commissioner, in February 2008 to ask for money. They claimed to have spent Euros 635m over the previous five years on CCS, and they went to say 'we expect that our companies in the aggregate will commit upwards of Euros 11.159bn over the next seven years.' But, stressing 'first mover risk' rather than 'first mover advantage', they said they faced 'unrecoverable costs...which cannot be fully justified to our companies' shareholders'. Therefore they needed 'transitional financial incentives' in the shape of a 'substantial' initial level of support.[6]

Chris Davies, a UK Liberal who was the European Parliament rapporteur on the directive to create a legal framework for CCS, said he was very conscious of the game the utilities were playing. 'I have yet to find any power generator without a hand sticking out and a begging bowl attached to it.'[7] Nonetheless, he believed that, in order to kick-start CCS, something quite big had to be dropped into the begging bowl. Bigger than the original Commission proposal that 60m allowances should be taken from the new entrants' reserve in the ETS, and be used to subsidize a dozen early CCS demonstration projects. The proposition was that not only would a CCS operator not have to buy an allowance for any tonne of CO_2 that was captured and stored, but he would also, for that same tonne of safely-stored CO_2, be given one or more ETS allowances that he could then sell.

Mr Davies won parliamentary support for increasing the 60m to as much as 500m allowances. But as part of the December 2008 agreement, EU governments decided that the subsidy should be 300m allowances, of which no more than 15 percent going to any one project. MEPs accepted this. The same

---

6   Letter dated 21 February 2008, see *www.zero-emissionplatform.eu*
7   Author interview, July 2008

agreement encouraged national governments to use a portion of their revenue from auctioning ETS allowances as a subsidy to CCS. With governments loath to make any immediate pledges of taxpayers' money for CCS, a future raid on the larder of ETS allowances seemed a very convenient solution. In theory, now there is legislation enshrining this subsidy in law, potential CCS operators can go to a bank and raise finance on the back of it. In practice, bankers will have to weigh carefully the future value of ETS permits as collateral for their loans, and they may not be reassured by the uncertain impact of the December 2008 deal on the ETS market.

Nevertheless this arrangement could mark the opening up of a new channel of funding for energy R&D in Europe. At their December 2008 summit EU leaders issued a declaration noting 'their willingness to use at least half' of ETS allowance auction revenue for climate control purposes, including R&D into low-carbon energy. Such a declaration is far from a binding commitment, but nor is it necessarily meaningless for the future.

CHAPTER 15

DOING WITHOUT

Negajoules represent the biggest energy source in Europe – ahead of oil, gas, coal and nuclear.

European Parliament, 2006

The EU has given itself a target to improve energy efficiency by 20 percent by 2020. But that does not mean an aim of using 20 percent less energy in *absolute* terms by 2020 – if it did, meeting it would almost automatically fulfil, and make redundant, the other target of cutting emissions by 20 percent. Instead, the energy efficiency goal is to save 20 percent of energy consumption *relative* to what the EU's energy is projected to be by that date if Europe just continued with its business as usual.

In other words, it is a pretty soft target. It differs from the 20 percent targets for cutting emissions and raising renewables in three ways. It is not binding. Its contribution is harder to gauge because it is measured not against a past base year but a future estimate. And its fulfilment depends on a wider range of actors, on the actions and reactions of virtually all of Europe's 500m citizens.

But progress in energy efficiency is very important because reduction in energy consumption, even if relative, will exert downward pressure on energy prices, and cut both imports and pollution – the three totemic goals of EU energy policy. Progress towards the energy efficiency target will also influence progress towards the other two targets. As regards the ETS, the higher the energy saving, the lower the demand to buy carbon permits and the lower the carbon price. The knock-on effect of that on, say, nuclear power may not be good. But the lower energy demand, the easier it becomes to meet it by renewable means.

But if the importance of energy efficiency is evident, the EU dimension is less obviously relevant to this aspect of energy policy than to other areas already discussed. It is axiomatic that design of the EU's internal energy market must be decided at the level of that market; it is only natural that countries seeking greater energy security should band together; and it is clear that global problems like climate change require the widest possible response, with a regional bloc of 27 countries merely a starting point. But energy saving is often seen as something done within the privacy of one's home or within the confines of one's state.

Brussels' usefulness, or otherwise, in energy saving policy is underrated, partly because energy saving or efficiency gets little attention in general. Deciding to save energy is not a process that brings EU member states into conflict with each other or creates press headlines. And actually saving energy, in the absence of some revolutionary gadget, is usually unglamorous. This is why, in the words of one EU official, 'there is a feeling that [energy saving] is so unconflictual that it will get done automatically, with a little help from oil prices.'[1]

Unfortunately, the rise in the oil prices since 2000 has not been that much help. Certainly overall energy use in the EU is fairly flat, amounting to 1,637m tonnes of oil equivalent in 2005 and showing no increase on 2004. And overall energy intensity – the amount of energy needed to generate a given unit of national wealth – continues to fall for the EU as a whole, to an average in 2005 of 208 kgs of oil equivalent for a 1,000 euros of gross domestic product (compared to 236 kgs of oil equivalent in 1995).[2]

But this average bridges an enormous gap. On the one hand, there is world-beating Denmark, whose energy intensity on the above measure is a miserly 114 kgs of oil equivalent, and at the other extreme, Bulgaria, a brand new EU member still with a Soviet industrial legacy, which uses energy 10 times more intensely than Denmark, at 1,582 kgs of oil equivalent. Yet even before Bulgaria's entry into the EU, the Commission was estimating in 2005 that the EU could save 'at least 20 percent of its present energy consumption in a cost-effective manner',

---

1   Author interview, 2007.
2   Eurostat, 2007.

which is where the 20 percent target sprung from.[3]

The business of energy saving is, moreover, complicated by the perverse effect of efficiency on demand – the more energy you save, the more you have to use for something else, and the more efficiently energy can be produced and the cheaper it becomes, the greater the incentive to use more of it. This perversity, known as the 'rebound effect', has been recognised for a long time, since indeed the 19th-century invention of the steam engine enormously improved energy efficiency but also increased energy consumption.

It is also what is happening with electricity today. Electricity consumption in the EU rose between 1999 and 2004 at 10.8 percent, almost exactly in line with GDP.[4] Now, there are reasons to favour a continuation of the historic trend of progressively electrifying the European economy into areas such as transport that might otherwise be hard to decarbonize. It could, for instance, enable electric cars to recharge with low carbon energy by plugging into a renewable or nuclear-generated grid. Therefore transport is a sector where an increase in *electricity* intensity might be good. But there are many examples of increased efficiencies in the generation or use of electricity that have simply stimulated consumers' appetite for more of the magic electrons.

On the generation side, there is the success of combined cycle gas turbines (CCGTs) bringing down the cost, and in most instances the price, of electricity in a way that encourages consumption. Far more numerous are the improved efficiencies in the amount of electricity used by household appliances such as refrigerators and washing machines. These efficiencies go hand in hand with a rise in household consumption of electricity, because of the increased penetration of appliances such as air conditioners, dishwashers and tumble driers. These developments constitute real welfare gains for people who can now afford to buy useful household goods that compared to the past, use relatively smaller amounts of electricity made relatively cheaper by CCGTs.

3   Green Paper on Energy efficiency, European Commission, 2005.
4   Joint Research Centre, 2007, *http://ies.jrc.ec.europa.eu*

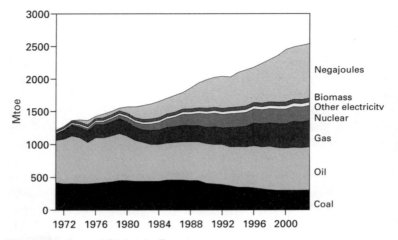

Figure 8: Energy Saving in Europe

Source: European Commission Green Paper on energy efficiency COM (2005) 265, p. 10

Less positive is the increase in standby electricity consumption from entertainment electronics, computer equipment and modern versions of traditional white goods that are fitted with special displays and microprocessors. Nor, in households, do these modern gadgets replace older ones as they would in businesses; older TVs are often shifted to children's bedrooms rather than thrown out. Newer appliances use less standby electricity. But the simple number of appliances with standby power mode continues to increase, and so therefore does overall consumption.

So is the energy conservationist on the hopeless treadmill encapsulated in the Red Queen's warning to Alice that in Wonderland 'you have to run as fast as you can just to stand still, if you want to get anywhere else you must run twice as fast as that'? Not quite. A recent UK study of the 'rebound effect' confirmed that the phenomenon certainly exists in both direct and indirect forms, and in total 'the evidence suggests that economy-wide rebound effects will be at least 10 percent and often higher of the energy saved.'[5] The direct rebound effect is

5   'The Rebound Effect', UK Energy Research Centre, 2007, pp. vii–xi.

where people use the money they save on energy to consume somewhat more of the same energy. This is likely to be higher in poorer countries or among poorer people because their demand for energy is less satiated. In developed countries, this same UK study suggests that the direct rebound effect for household heating and cooling and for personal transport 'is likely to be less than 30 percent and may be closer to 10 percent for transport'. The direct rebound effect may be somewhat larger when producers, rather than consumers, adopt energy efficiency technologies such as the steam engine in the 19th century or the electric engine in the 20th, because producers' appetite for energy will not be limited by their personal needs. An indirect rebound effect can occur where people (usually richer people) use the money they save on one kind of energy, say, electricity for heating and lighting, and spend it on another, say, kerosene to jet them away on another holiday.

Yet none of these effects really matters as long as energy efficiency improvements do not so stimulate demand that overall energy use actually increases. Energy economists term this counter-productive effect 'backfire', and it did occur when steam and electric engines arrived on the scene. The UK study underlines 'there is no a priori reason to believe that "backfire" is an inevitable outcome in all cases.' Even where an energy saving technology results in a rebound effect of 30 percent, 70 percent of the energy saving is still preserved. But this UK report suggests caution in estimating the actual energy savings from energy efficiency technology. It is also a reminder of the importance of carbon and energy pricing in reducing rebound effects, by keeping the cost of energy constant while efficiency in producing it improves. And with the issue of energy pricing, on which the EU has some agreements, the EU enters the picture.

What role, then, for Brussels and policy-making at the EU level? There is some ground to think that EU institutions – meaning the Commission and Parliament – are more inclined towards action on energy saving than national governments. Although Brussels is often considered as putting producers' interests above those of consumers (the most famous example being the Common Agricultural Policy), this is not so evident in energy. Indeed, arguably, the locomotive that has driven

EU energy policy forward for some time has been Brussels' competition directorate with its anti-trust investigations, usually launched on complaints from energy users. By contrast, national energy ministries in the member states tend to be more influenced by energy producers, which, everything else being equal, are interested in customers buying more not less of their product. 'Because of "agency capture" by energy interests of the department of trade and industry', claims Andrew Warren of the UK Association for the Conservation of Energy, 'almost nothing would happen in UK energy saving if it were not for Brussels.'[6]

Of the two main instruments available to encourage people to do the unnatural thing of saving energy – regulation and taxation – the EU has so far overwhelmingly relied on regulation.

Products

The EU's main recent actions are the 2003 energy labelling directive and the 2005 Eco-design directive. The first required manufacturers to put clear information about energy use on labels on their products. It also set some minimum energy efficiency standards for products, but these did not go far enough. So the Eco-design directive was passed to set energy efficiency requirements for a wide range of consumer goods ranging from water heaters to TV set up boxes. In July 2008, the Commission proposed extending this directive to all energy-related products – those that do not consume energy during use but have an indirect impact on energy consumption, such as (hot) water-devices or windows.

The first concrete measure taken under the Eco-design directive was agreement in July 2008 to cut the electricity consumption of standby devices in offices and homes by nearly 75 percent by 2020. Another decision under this directive is to phase out incandescent light bulbs by the end of 2012, replacing them with more efficient fluorescent light bulbs.

---

6   Author interview, 2008.

Transport

In this sector, the EU has a role of Union-wide dimension, in the air and on the ground, which has already been explored in Chapter 10. Legislation is agreed to put the emissions from all aircraft using EU airports (including non-EU airlines) into the ETS in 2012. At the same time, the Commission is helping Eurocontrol to try to create a coordinated air traffic system with the Single European Sky programme that could, by reducing aerial congestion and stacking over Europe's airports, cut aviation fuel consumption by an estimated 11 percent.

Partly because cars, or their drivers, are not, like airlines, conveniently organized into fleets that could be slotted into the ETS, legislation has been directly imposed on the car industry to reduce vehicle emissions. The approach is similar to the Corporate Automobile Fuel Efficiency (Café) standards in the US, only tougher.

Buildings

Some 40 percent of all the energy in the EU is used in buildings and the potential savings on this energy is considerable, as much as 28 percent by 2020, according to Commission estimates. In 2002, the energy performance in buildings directive was passed, though it only came into force in 2006. Even then it only required member states to have an energy performance standard of their own for large new buildings of more than 1,000 square metres or similarly sized existing buildings undergoing renovation. This, however, was progress because most of the new Central European states had no such standard, until their entry into the EU. The most important change would be to require the retrofitting of energy saving equipment to existing buildings not undergoing major renovation. Imposing standards on buildings does not deal with the fact the stock of buildings takes far longer to 'turn over' than that of products.

Public procurement

The EU institutions and the central governments of member states have amended the legislation on their public procurement – amounting to 16 percent of EU gdp – to make energy efficiency a criterion for choice when they buy goods and services. In this way Brussels has adapted one of its more powerful internal market instruments to the cause of energy efficiency.

This public procurement legislation has been used to thwart the natural tendency of national governments to award contracts to their own national companies, so segmenting the market. The legislation requires all government contracts over a certain value to be advertised electronically across the EU. In the past, it has generally required that, if all conditions such as quality and safety are equal, contracts should go to the cheapest bidder, as a safeguard against protectionism and corruption (inflated price contracts can conceal kick-backs). Such an approach can, however, discourage innovation and new energy-saving technology that is often, at least in the short run, more expensive than what it replaces.

In 2004, 'green public procurement' guidelines were introduced to encourage local authorities to factor into the costing the life-cycle costs (such as emissions escaping during the production process or the running cost of a building) of products they were tendering for. However, only a few member states – only seven according to a 2006 study – appear to have embraced this. The Commission announced in mid-2008 further efforts to promote green public procurement.

Trade

So far trade policy has not figured much in the EU's quest for energy saving or emission reduction, with the minor exception of the 2007 EU – US agreement to highlight the energy efficiency saving merits or demerits of office technology, through labelling. But trade policy will play an increasing role in policing energy inefficient imports that would undermine product standards in the EU. Sometimes external trade issues have pushed internal regulation along.

An example of this has occurred in the greater light bulb switchover, which will eventually see Thomas Edison's incandescent light bulbs phased out and replaced by fluorescent light bulbs. The EU's Eco-design of Energy-using Products of 2005 gave the Commission the choice of either accepting industry promises of self-regulation or tabling mandatory legislation. Somewhat counter-intuitively, the industry represented by the European Lamp Companies Federation (ELC) quickly opted to have compulsion imposed on it, and crucially on its foreign competitors. ELC's secretary general, Gerald Strickland, said, 'we decided that the voluntary route would offer no control over, or sanction on, importers continuing to undermine our efficiency efforts with inefficient products.'[7] Of course the great majority of new energy-saving bulbs will come from China, mainly from ELC company subsidiaries there.

The hard part of trade policy will be to ensure that legitimate policing of imports for observance of EU standards stops short of protectionism. Moreover, it would probably overstrain trade policy if, in a bid to equalise carbon controls, the EU were to start evaluating the emissions not just of foreign products, but also of the process by which these products were made. For instance, what if import into the EU of fluorescent light bulbs from China were blocked or penalized because the EU judged there was insufficient control on the carbon emitted during the manufacture of those bulbs?

Tax

While trade is an area of coming involvement for Europe's energy policy makers, energy taxation is one of EU member states' older battlefields – and one that may soon have to be revisited. EU governments accept the need for some harmonization of indirect tax rates on motor fuels in order to prevent serious distortions in the markets for these valuable, mobile and

---

7   Author interview, April 2008

generally highly taxed commodities. So there is an EU system of minimum tax rates on petrol, diesel and other mineral oils, and the 2003 Energy Taxation Directive extended these floor rates to other energy sources such as coal, gas and electricity.

But the EU only taxes energy when it is used as fuel or for heating, and not as raw materials in industrial processes, or as input in the making of other energy products (in refineries) or even as inputs for electricity generation. The European Commission tried to remedy this back in 1991, in the run-up to the United Nations conference in Rio de Janeiro that put climate change on the political map. It proposed a wide-ranging energy tax, calculated on both energy content and on proportion of carbon emissions. The proposal foundered, mainly on opposition from the UK (on the political grounds of fiscal sovereignty) and from Spain (arguing such a tax would cramp its development).

Yet, very tentatively, the Commission is trying to return to the issue. In spring 2007 it ventured the thought in a green paper that 'the explicit identification of an environmental element in the minimum levels of taxation (differentiating between greenhouse gas and non-greenhouse gas emissions) would enable energy taxation to complement other market-based instruments at EU level.'[8] And there is a good argument for doing so, particularly to reach parts of the economy that the emissions trading scheme (ETS) itself cannot easily reach.

As the main framework for controlling climate change, the quantity allocation method is peculiarly apt, for both technical and political reasons. Technically, because the science has given us a ballpark figure of the amount of greenhouse gases we want to take out of the atmosphere, or rather the level of so many parts per million that we want to limit greenhouse gases to. So the ETS works on the basis of the authorities setting the quantity of carbon to be reduced, and letting the market set the price of doing that.

Politically, too, this has several advantages. One is simply that while the ETS is effectively a tax, it is not called one. This enables it to be swallowed in the EU context by the UK, and

---

8   Green Paper on market-based instruments for environment and related policy purposes, COM (2007) 140 final, p. 8.

perhaps one day in the context of an extended post-Kyoto system by the US, a country even more jealous of its fiscal sovereignty. The other political plus is the opportunity to smoothly phase in schemes like the ETS by initially giving out some pollution permits for free, though eventually most or all permits must be auctioned if they are to have a real cost that changes the polluter's behaviour.

However, an ETS involves calculating, and controlling, individual permit levels for individual polluters or energy users, and this becomes quite impractical for the likes of small businesses, households and car drivers. So there is a case for reviving the idea of a carbon tax (from which sectors/companies covered by the ETS might be exempt). It would also keep the cost of energy services constant despite any efficiency improvements, and would therefore minimize 'rebound' effects.

Returning to an old theme it first raised in 1993, the Commission likened such a carbon tax to 'an environmental tax reform shifting the tax burden from welfare-negative taxes (e.g. on labour) to welfare-positive taxes (e.g. on environmentally damaging activities such as resource use or pollution)', and therefore producing 'a win-win option to address environmental and employment issues'.[9] Taxing environmentally damaging consumption might also help governments replace revenue from taxes that, in the era of globalization, are getting harder to levy on capital. Environmental taxes would be regressive (because higher energy charges would take more out of the pocket of the poor than of the rich). But this effect could be offset if governments cut labour and social security charges at the lower end of the income scale.

While an EU-wide carbon tax complementing the ETS would have many advantages, the requirement of unanimity among the EU's 27 governments on tax issues is not one of them. Some governments, notably the UK and France, are showing interest in reshaping EU level taxation in a green way. At their March 2008 summit, EU leaders invited the Commission to 'examine areas where economic instruments, including VAT rates, can

---

9   White Paper on Growth, Competitiveness and Employment, COM (93) 700 Chapter 10.

have a role to play to increase the use of energy-efficient goods and energy-saving materials'. But reducing VAT on some energy efficient products or services – which might get the required government unanimity – would not create the widespread change that an economy-wide carbon tax would bring.

National action

Part of the 20 percent efficiency improvement – a saving only compared to what energy use would otherwise be – is supposed to come from national programmes. This is in addition to whatever energy and emissions saving are made as a result of the ETS or other EU-wide measures. Under the 2006 Energy End-Use and Energy Services directive (which like every directive, of course, had to have governments' agreement), member states have been required to file national strategies on how they planned to achieve a (non-binding) goal of reducing energy consumption by nine percent over nine years.

The lackadaisical way in which many member states have implemented this directive gives the impression that they do not care much about energy saving, or, if they do, that they do not regard the EU as very relevant to this task. The National Energy Efficiency Actions Plans (NEEAPs) were all supposed to be filed by July 2007. But Commission had to chivvy governments with threats of court action, and it was July 2008 before the last (from Greece) of the 27 plans straggled in.

These plans may be of some use to the Commission as an information exercise of what is or what is not being done at national level. This directive could, in the words of a Commission official, 'provide us with the means to look into member states' backyards in terms of energy saving and to see what more could be done at EU level'.[10] But in mid-2008 the Commission still had 16 infringement proceedings against member states for failing to transpose the directive, correctly or at all, onto their statute books. For the most part the national plans are distinctly unimpressive in their ambition, though the Commission has

---

10 Author interview, 2008

been gentle in its public assessment of them. In January 2008, it reported on the 17 plans it had received by then. The nearest it got to any criticism was to say that, while 'several present comprehensive strategies and plans are likely to deliver savings beyond the required nine percent, many seem to present a business-as-usual approach.[11] This was hardly the naming-and-shaming tactics that Brussels uses against member states that drag their feet on single market legislation.

Yet, while the Commission should get tougher in prodding member states into energy conservation, decisions on what measures to take must very often be made at national level, taking advantage of simpler local procedures and better local knowledge. Acts of individual leadership, such as the decision of Ken Livingstone to run for mayor of London virtually on the single issue of a traffic congestion charge for the UK capital and to carry it through, are hard to envisage in the more complex, collective context of EU policy-making. The same could be said of the Irish government's decision to place a green tax on plastic shopping bags. The wisdom of devolving decisions downwards wherever possible, and only taking them at the EU level where necessary, got formal recognition when the subsidiarity principle was enshrined in the 1992 Maastricht Treaty.

Despite the increasing degree of compulsion applied to it, energy saving remains something of a cultural issue (and as regards climate change, a moral issue in the sense of a moral obligation to future generations). Attitudes towards energy conservation will therefore evolve in the way they have towards smoking. There is no collective European conscience about energy saving, as there is against the death penalty that every European government has repealed. Some countries care more than others about energy saving. It would be tempting to generalize that northern Europe cares more than southern Europe, which generally sees itself as more in a catch-up phase of energy-driven development. But inside the northern belt of EU countries, indeed inside Germany, there is an odd contradiction. Germans are model recyclers of household and consumer product waste and have led the way in

---

11 'Moving Forward Together on Energy Efficiency', Commission communication, COM (2008) 11 final, p. 12.

renewable energy. But they are apparently addicted to conspicuous energy consumption in the shape of big, therefore heavy, and therefore CO_2-emitting, cars – and to the freedom to drive these cars as fast as they like on their autobahns.

Kicking the energy waste habit is likely to evolve unevenly across Europe. Only gradually may climate change concerns permeate into a common consciousness about energy wastefulness. Ironically, one factor promoting a common consciousness in general among Europeans has been their ability to fly all over their continent on budget airlines, a phenomenon created by EU aviation liberalization but which now, awkwardly, adds to global warming.

However, because of the urgency of climate change, regulation of energy waste will have to run ahead of social attitudes to it. This will be tricky for politicians at the EU and national level. They will dare not get too far ahead of voters. Leadership in energy policy is especially difficult, because energy policy changes entail lifestyle changes, and usually an element of personal sacrifice.

CHAPTER 16

CONCLUSION

The last few chapters have not been a stellar advertisement for energy policy at the EU level. EU research in nuclear power has focused too much on the ever-distant prospect of nuclear fusion and too little on the here-and-now problem of disposing of nuclear fission's radioactive waste. EU research has been inadequately funded, particularly in view of the need to get carbon capture and storage demonstration technology going in time to make a difference to the climate. The Commission has preached the merits of energy saving to member states, but has been unable to get their agreement to an EU-wide energy/carbon tax that would now be a useful complement to the ETS. And the EU has yet to establish itself in the eyes of its member states as a direct provider of security for their energy imports.

Yet the Commission's hard slog to integrate Europe's energy market is paying off, if not in terms of lower gas and electricity prices (which fluctuate up and down with the oil price), then in terms of indirect energy security. Better inter-connections between the EU's 27 national electricity and gas markets will increase the ability to move supplies around in time of shortage, and improve the Union's resilience to withstand supply shocks. The jury is still out on how effective the EU's climate change policies will prove in practice for cutting emissions. The proposed post-2012 reform of the ETS is an improvement, but there are still doubts about the design of the renewables and biofuels parts of the new climate change programme. However, the principle and value of collective EU action on global warming is, rightly, unquestioned throughout Europe.

This book has argued that the EU offers high potential for collective action in energy (especially on climate change), and has complained where this potential has not been realized (especially in nuclear power). It has also highlighted a few ways in which the EU tries to prevent member states from making bad policy trade-offs, especially on energy pricing. Artificially low energy prices interrupt the play of supply and demand, depress incentives for new domestic supply or generation, encourage energy waste, and thereby increase dependence on imports and exacerbating emissions. Governments do not like the Commission taking them to court for interfering with liberalization by putting a ceiling on energy prices to placate voters. But the Commission is trying to save governments from their own short-sightedness. The right way to deal with fuel poverty is to target subsidies to those in real need of help to pay their energy bills.

To some extent, EU member countries gain security by simply *being in* a Union rather than necessarily *doing anything as* a Union. The EU gains as a whole from the diversity of its member countries' energy mixes, though some countries gain more than others. France, for instance, is a net provider (in the absence of any reactor accident) of energy security with its nuclear-generated exports to neighbours such as Italy, which with its high dependence on gas imports is a net consumer of energy security. But a Union is not an energy union unless it has networks linking its national markets commercially as well as physically. Hence the long battle about network unbundling.

The fact that energy could have become such a major policy activity without becoming a formal treaty competence of the Union must be a tribute to the value placed on energy policy making at EU level. It is also something of a puzzle to political scientists. They like to debate what, and who, drives on this strange beast of a European Union, and the even stranger prospect of an energy policy that has advanced without a specific legal base to it. Is it 'inter-governmentalism' in which governments predominantly take the initiatives and the decisions? Or is it more what some political scientists call 'multi-level governance', with EU institutions, principally the Commission, and pan-European interest groups playing the major role?[1]

The answer is a bit of both, at different times and according

---

1   Jon Birger Skjaerseth and Jorgen Wettestad, *EU Emissions Trading: Initiation, Decision-making and Implementation*, Ashgate, London, 2008, pp.15–16.

to events. In one of the few books on EU energy policy, Janne Haaland Matlary was able to write in 1997 that 'since this [energy] is a new area where the EU institutions enjoy no formal competence, we can safely assume that the member governments will remain at the forefront of the process [of policy-making].'[2] This was plausible in 1997, when France and Germany were still putting a brake on the beginnings of energy liberalization, when the energy security issue was temporarily dormant, and before the implications of the Kyoto Protocol signed that year became clear.

But the follow-on to Kyoto produced a very different dynamic in which the Commission took the lead. According to two other Scandinavian political scientists, Jon Birger Skjaerseth and Jorgen Wettestad, the EU had been suspicious during the Kyoto negotiations of the US-promoted idea of emissions trading; it was seen a soft option, even a let-out, and the EU was much influenced by the green groups' slogan at the time that 'trading pollution is no solution'. But shortly after Kyoto, there happened to be a staff turnover in the Commission's environment directorate in which officials 'who favoured command-and-control were replaced by economists who favoured economic policy instruments'.[3] These environmental economists took to the idea of emissions trading, even though most member states were opposed to it. After the US formally abandoned Kyoto in 2001 – a move that shocked the EU but also galvanised it into a leadership role – Brussels' environmental economists turned emissions trading into what was seen as the main instrument for saving the protocol.

The ETS was a good example of how EU policy advances where, first, the European dimension is seen to be relevant and, second, where the EU, especially its executive arm, the Commission, is seen to be useful. With these watchwords of relevance and usefulness in mind, it is time for the EU to start ranking priorities in energy policy. The Commission likes to claim equal weight for all three goals of its energy policy. But

---

2   Janne Haaland Matlary, *Energy policy in the European Union*, Macmillan, London, 1997, pp.2–3.
3   Skjaerseth and Wettestad, p.99.

where they conflict, or compete for political attention and effort, one goal has to take precedence over the others, and that should be combating climate change.

Internal market

Priority no longer needs to be given to achieving a real internal market in energy. To say this is not to downplay the progress that has been made. Nor is to argue that because liberalization is an old agenda it is a past agenda. Rather, liberalization is a battle that the regulators and anti-trust officials, in Brussels and national capitals, must still wage daily so as not to let incumbent companies fix energy markets in the name of energy security, or government officials fix energy prices in the name of social regulation. Moreover, the extraordinary amount of time and effort that the Commission has put into unbundling networks over the past dozen years has not been wasted. These networks are Europe's energy arteries, and clearing out discrimination and protectionism is vital. The Commission is correct that its work to create a more competitive and integrated internal market does generally serve the cause of climate change control (by removing market distortions that would blunt the correct carbon-pricing signals for new investment) and of energy security (by giving Europe inner resilience against outside supply interruptions).

The Commission was also right to declare a victory of sorts, and to strike a compromise with the die-hard opponents of ownership unbundling. The latter were forced into admitting that some further legislative reform was needed to make Europe's energy network operators more independent and give them more incentives to extend grids and pipelines. It had therefore become pointless to go on crusading against the general business model of vertical integration, when specific market abuse could still be pursued with anti-trust law. The compromise was less tidy than the Commission's original blueprint. Yet the EU's executive body could take satisfaction, from an international perspective, at having engineered more of a standard market design across 27 widely-differing nation states than the US federal authorities have ever achieved, at least in electricity, across America's 50 states.

A fresh package of legislation can now go ahead. In addition to reinforcing unbundling, this can create new pan-EU bodies of regulators and transmission system operators whose top concerns should be greater transparency as well as encouraging new entrants and more grid connections to link national markets. This would improve energy security.

Energy security

As the Commission rightly said in its November 2008 'Strategic Energy Review', 'security of supply is more and more seen as a public good deserving a closer attention from the European Union.'[4] This was the Commission's understated way of alluding to east European states' clamour for an energy security strategy and their threat of a collective veto on other energy and climate change policies unless they got such a security strategy.

Energy security has never been left solely to the private sector at the national level, and cannot be at the European level. It is illusory to imagine that Europe's bigger energy companies, as they evolve into pan-European concerns, will or can automatically assume 'European' responsibilities for energy security. True, now that Electricité de France has bought, with its British Energy acquisition, most of the UK's existing nuclear reactors, and plans to build four more, the French company will have to 'think British or European' as well as 'French'. Likewise, Germany's Eon, which supplies almost all of Hungary's gas, now has to 'think Hungarian' as well as 'German or European'. But these multinationals will not voluntarily add to their costs in terms of, say, extra reserve generating capacity or interconnections or gas storage, unless they are told what to do by governments and regulators at the national level, and increasingly at the EU level.

Belatedly and piecemeal, the Commission is coming up with a strategy that has a shorter-term focus as well as relying on

---

4 *http://ec.europa.eu/energy/strategies/2008/2008/2008_11_ser2_en.htm*

renewable energy to wean Europe off imported fossil fuels over the long term. The November 2008 'Strategic Energy Review' contained plans to encourage filling in the most glaring gaps – such as between the Baltic states and the rest of the EU – in energy networks. The Commission also floated the idea of giving the EU, and therefore itself, the money and power to influence decisions on infrastructure planning, an area where the EU currently has no legal competence. Another focus of the strategic review was a revision of the EU's absurd 2004 Gas Security Directive, as well as updating of oil stocks legislation.

But the main reason why the EU wants to do more to assure its energy security through its own means is its discomfort in relying ever more on Russia, for oil, coal and uranium but especially for gas. Russia constitutes the EU's most important but most difficult energy partner. It is the most important, because Russia will remain the largest single supplier of gas to the whole Union as well as currently the sole gas supplier to some land-locked EU states in central Europe. It is the most difficult, because the gas comes not just from an 'ordinary' state-controlled monopoly supplier (of which there are many in the gas business), but from the world's largest vertically integrated gas monopoly, Gazprom, which carries political baggage like no other commercial company. Gazprom is effectively controlled by a government with which, for reasons of history and geography, many EU states have a broad range of political differences, past and present. Further geo-political complications arise from Gazprom's control of most of the outlets for central Asian gas and from Gazprom's uncertain relations with the transit countries of Ukraine and Belarus.

Wouldn't it be nice if, instead of Gazprom with its legal export pipeline monopoly, there were five or six Russian private sector gas companies – none with a foot in the Kremlin – vigorously competing with each other to supply the EU and with equal access to Russia's pipelines to do so? After all, the EU was able to force Norway to give up its GFU single selling organisation for Norwegian gas into the EU, which Brussels deemed to be an illegal cartel. But this was because, in joining the European Economic Area (EEA), Norway had consented to abide by EU single market and competition rules. It is simply day-dreaming to imagine Moscow might agree to anything similar or even to real reciprocity in an energy agreement with the EU.

Perhaps the EU and Russia will negotiate a new Partnership and Cooperation Agreement. But the EU is unlikely ever to persuade Russia to give EU companies as free access to upstream exploration or to midstream pipelines as Russian companies have, at least in theory, to establish themselves in the EU's downstream energy sector. President Dmitri Medvedev, who was chairman of Gazprom before he succeeded Vladimir Putin as president, is hardly the man to reverse his predecessor's reinforcement of Gazprom's domestic monopoly. But while the EU cannot force open the Russian energy market, it can apply the full force of its competition policy to Gazprom's activities in the EU market.

It is therefore important not to compromise the principles of that policy. The Commission should therefore tread carefully in advancing its suggestion of a 'Caspian Development Corporation' to be a block purchaser of Caspian gas. The aim would be to encourage Caspian gas producers to sell to Europe, by grouping all EU purchases of Caspian gas into one big contract that might be underwritten by EU governments through the European Investment Bank. Such a scheme would need an exemption from anti-trust rules that might be excused on grounds of providing a 'pro-competitive' alternative to Russian gas. But allowing an EU buyers' cartel while, for instance, disallowing a Norwegian sellers' cartel would appear to bend the rules and condone double standards in a way that Gazprom might exploit.

In summary, the EU as a whole can do more to assure its energy security. And it must do more. Otherwise it risks a revolt from its newer members in central and Eastern Europe, angry that Union membership has given them an energy market policy they feel is ill-suited to their needs, climate change policies they regard as too expensive, and an energy security blanket that is nothing like as warm as they had expected. At the same time, these states should be wary of what they wish for. The way to a more effective EU energy security policy might lie through some surrender of national control over countries' energy mix. Would Poland really want to give Brussels and its EU partners a *droit de regard* over its power generators' near-total reliance on dirty coal?

Climate change

Climate change will dominate Europe's energy agenda. That is certain, simply because of the momentum from past commitments and current negotiations. The EU still has to meet its Kyoto commitment of an eight percent emission reduction in the 2008–12 period. The wide-ranging redesign of its climate change policies for the 2013–20 period still leaves the detail of many issues to be dealt with. These include tariffs and sustainability criteria for biofuels, carbon leakage and carbon equalization measures, and maybe the incorporation of shipping as well aviation into Europe's emission trading scheme. The EU will be at the heart of the process, because all member states agree solutions can only be found on the EU level. The Commission's role will be indispensable. The complexity of Europe's climate change response requires a policy architect of the Commission's (eventually acquired) expertise in this field. The sensitivity of sharing the burden between states needs a referee of the Commission's institutional neutrality.

It is also the case that climate change *should* dominate energy policy, because of the time urgency in tackling the steady accumulation of greenhouse gases. Of less intrinsic importance, except to European self-esteem, is whether the EU continues to dominate the international climate change agenda and negotiations in the way that it has in recent years. Indeed Europe may have to cede the baton of leadership to the US. Until President George W. Bush pulled it out of Kyoto, the US was very influential in the climate change talks – and it will be again, as it returns to the negotiating table under President Barack Obama.

This will be to the good, if US influence is the result of Mr Obama's persuasiveness in getting his country and the world to adopt ambitious emission reductions. However, the US may be influential simply because its relative lack of ambition in emission reduction may suit many countries during the expected economic downturn of 2009–10. EU targets, conceived in a time of buoyant growth, look ambitious in a recession, and may prove to be set a bit too high to give Europe bargaining leverage. For example, if no other country in the UN climate change negotiations is willing to make reductions even up to the equivalent of the EU's unilateral cut of 20 percent by 2020, what extra negotiating mileage does the EU get in proclaiming readiness to go to a 30 percent cut if others match it?

However, it would be a tragedy if a temporary economic slowdown were used as a pretext to weaken measures coming into effect from 2013 onward. For if the EU finds by 2020 that its policies have not been up to the job of serious emission reduction, it cannot try another solution and hope that all the parameters of the global warming problem will be unchanged. We know they will be worse.

This is why all measures need to be tried, using belt and braces. The emissions trading scheme has, rightfully, been identified as the key instrument. It is a market-oriented system that offers flexibility across sectors and across countries, and considerable sources of allowance auction revenue for governments. It has economic elegance and political convenience. But a lot is being asked of it: a 21% emission reduction (over 2005–20) from the sectors covered by the emissions trading scheme (ETS). This is still a young and relatively untried and untested market.

As insurance against the ETS failing to perform, perhaps more reduction should have been demanded of sectors outside the ETS than the ten percent cut Brussels has proposed for them over the same 2005–20 period. The stated rationale for the differential cut is that reductions are relatively easy to make in the electricity industry (in the ETS) and harder to obtain in agriculture, transport and services (non-ETS). In these latter sectors, nonetheless, the EU should show more ambition in saving energy and emissions.

In the redesign of EU climate change policy, many economists, and some governments, have complained that the inclusion of sub-targets on minimum renewables and biofuels levels is sub-optimal. Relying on the single target of emission reduction would, they say, have been more efficient. It probably would. And there is much to criticise in one aspect of renewable policy:

the restrictions on green power trading introduced to protect certain member states' lavish support schemes. Should Germany's solar power feed-in tariff – set so high that 86 percent of total EU solar power capacity is now installed in that one country – really be sheltered from competition that might lower this irrationally generous subsidy?

But what such economists fail to factor in is the sense of urgency that has to accompany climate change policy. Every profession has its blind spot. Political scientists are prone to over-theorize. When they look at the EU, some of them seem unable to grasp that the success of institutions lies in their practical relevance and usefulness – in short, 'what works works'. For their part, economists zero in on costs, efficiencies and ways of getting maximum output for minimum input. But this approach, valuable in tackling static situations, can have limitations in tackling climate change because it is such a dynamic problem, and we are late joining battle against it. So when analysing Europe's planned journey towards low-carbon energy, some economists focus too much on the efficiency, or economic comfort level, of travel rather than on the urgent need to get to the destination, almost by whatever means. This is why, in addition to the ETS, other measures – some outcome targets for renewables and biofuels as well as subsidies for putting carbon underground – are needed to ensure success.

In normal peacetime, policy failure is not catastrophic. If a policy fails, we can also ways demand that our governments re-do it. But combating climate change is more like war: you don't have time to return to the drawing board.

INDEX